SOTHIS PRESS

Sothis Press aims to completely reimagine the public domain book with an emphasis on universal access to captivatingly designed print editions. We believe that technology will enable every single book ever published to be perpetually accessible in beautiful print form. Sothis is beauty in ubiquity.

www.sothis.press
1 Waterhouse Square, London, ECIN 2ST

A Handbook of Siberia and Arctic Russia

General

G. B. Naval Intelligence Division

SOTHIS PRESS

ISBN 978-1-0365-1950-6

A Handbook of Siberia and Arctic Russia
is in the public domain.

This edition is published by
Sothis Press © 2023. The company
is registered in England and Wales.
The Company Number is 14900389.

I. D. 1207

A HANDBOOK OF SIBERIA AND ARCTIC RUSSIA

VOLUME I

GENERAL

Compiled by the Geographical Section of the Naval Intelligence Division, Naval Staff, Admiralty

LONDON:
PUBLISHED BY HIS MAJESTY'S STATIONERY OFFICE.
To be purchased through any Bookseller or directly from
H.M. STATIONERY OFFICE at the following addresses:
IMPERIAL HOUSE, KINGSWAY, LONDON, W.C. 2, and
28 ABINGDON STREET, LONDON, S.W. 1;
37 PETER STREET, MANCHESTER;
1 ST. ANDREW'S CRESCENT, CARDIFF;
23 FORTH STREET, EDINBURGH;
or from E. PONSONBY, LTD., 116 GRAFTON STREET, DUBLIN.

Price 7s. 6d. net

Printed under the authority of
HIS MAJESTY'S STATIONERY OFFICE
By FREDERICK HALL at the University Press, Oxford.

NOTE

The region covered in this Handbook includes besides Siberia proper, that part of European Russia, excluding Finland, which drains to the Arctic Ocean, and the northern part of the Central Asian steppes. The administrative boundaries of Siberia against European Russia and the Steppe provinces have been ignored, except in certain statistical matter, because they follow arbitrary lines through some of the most densely populated parts of Asiatic Russia.

The present volume deals with general matters. The two succeeding volumes deal in detail respectively with western Siberia, including Arctic Russia, and eastern Siberia.

Recent information about Siberia, even before the outbreak of war, was difficult to obtain. Of the remoter parts little is known. The volumes are as complete as possible up to 1914 and a few changes since that date have been noted. No attempt, however, has been made to give any account of the social, and political and economic conditions which are the outcome of the Russian revolution of 1917.

Russian statistics have never been very trustworthy. As regards Siberia, when given separately from Russia and Central Asia, they are seldom of recent date. Such statistics as are obtainable are given in the chapters to which they refer. All figures relating to population must be accepted with caution.

The Admiralty will be glad to receive corrections and additions.

CONTENTS

А а *а*	*a*	
Б *Б* *Б* *б* *б*	*b*	
В *в*	*v*	
Г *Г* *г*	*g*	*h* in foreign works
Д *Д* *д* *д* *д*	*d*	
Е е	*e*	*Ye* when initial
Ж *Ж* *ж*	*zh*	
З *з*	*z*	
И и *И*	*i*	
І *І* *і*	*i*	
К к	*k*	
Л *Л* *л*	*l*	
М м *М*	*m*	
Н н *Н*	*n*	
О о	*o*	
П п *П*	*p*	
Р р *Р*	*r*	
С с	*s*	
Т т *Т* *Т* *т* *т*	*t*	
У у	*u*	

Ф ф *Ф Ф Ф*	*f*
Х х *Х*	*kh*
Ц ц *Ц*	*ts*
Ч ч *Ч*	*ch*
Ш ш Ш *Ш*	*sh*
Щ щ *Щ*	*shch*
Ъ ъ *Ъ ъ*	mute
Ы ы *Ы*	*i*
Ь ь *ь*	mute
Ѣ ѣ ıь *Ѣ*	*ye*
Э э	*e*
Ю IO ю	*yu*
Я я	*ya*
Ѳ ѳ	*th*
Ѵ ѵ *Ѵ*	*i*
Й й *Й*	*i*

The combinations ЫИ and IИ are transliterated *i*

NOTE

In the above table the principal forms of the letters of the Russian alphabet, printed and cursive, which occur in official Russian maps are given. In actual practice little or no distinction is made between printed and cursive forms, and consequently they have not been separated in this table. In the case of each letter only the commonest form of the small type is given, but in nearly all cases any form of the capital type, reduced in size, may be used.

This system is the same as that used by the Admiralty except as regards ц, which the Admiralty transliterate tz instead of ts. The War Office system differs from the one adopted in this book by transliterating ж by j instead of zh. The sound of this letter is represented by the French j, which is the equivalent of zh in English. The Royal Geographical Society's system is the same as that of the Admiralty.

All proper names have been transliterated from official Russian maps. In the frequent cases of disagreement between different maps the 40-verst, or failing that the 100-verst map, has been preferred. Only words indicating geographical features as bay, lake, &c., have been translated. Russian words capable of translation which form the whole or part of a proper name have not been translated, but transcribed into Roman characters, e. g. Byeli, Nizhne. The only exceptions to these rules are in the case of names in common English usage such as White Sea, New Siberia Islands, &c.; and names that were originally English or French or of other foreign languages and have been adopted by the Russians, as de Castries Bay, Nordenskjöld Archipelago, Jeannette Islands, Valentine Bay, &c.

In order to simplify reference to Russian maps the adjectival endings showing gender of all Russian names have been retained. The result of this is a difference in the versions of the same name applied respectively to a bay (feminine in Russian), a village (masculine), or a church village (neuter).

NOTE

MAPS OF SIBERIA AND ARCTIC RUSSIA

The only official map of Siberia which covers the whole country is the *Map of the Oceans, Seas, Rivers and Lakes of Asiatic Russia and Adjoining Lands*, scale 100 versts to an inch (1 : 4,200,000), published by the Ministry of Ways and Communications, 1905. This map is in four sheets. It is coloured to show the drainage areas of the chief rivers. No attempt is made to show relief. There is much detail in relation to rivers, towns, villages, and tracks, but the map is untrustworthy in many parts. Moreover it is badly printed from worn type and somewhat illegible.

The topographical section of the Russian General Staff publishes a *Map of the Frontier Regions of Asiatic Russia*, scale 10 versts to an inch (1 : 420,000). There are twenty sheets in all of various dates from 1886 to 1911. The country covered is from about lat. 58° N. to northern Mongolia, but in western Siberia the sheets go north almost to the Ob delta. All Russian Central Asia is covered. The map shows relief by hill shading, but a great deal of it appears to be imaginary. Beyond the better known districts along the chief rivers and the railways this map cannot be relied upon. Roads and tracks are shown and a great many names are marked. Rivers are clearly indicated in blue. The map is generally legible. Little reliance can be placed on the altitudes, given in feet.

The Ministry of Ways and Communications also published in 1911 a *Map of Communications of Asiatic Russia*, scale 100 versts to an inch (1 : 4,200,000) in three sheets. The map shows roads, railways and navigable rivers, as well as telegraph lines. There are rough indications of topographical relief. The map is clear and legible if somewhat diagrammatic, but like other maps of Siberia is far from accurate in many parts.

The Ministry of Ways and Communications also publishes a *Map of Communications of Asiatic Russia* in one sheet on a scale of 300 versts to an inch (1 : 12,600,000). The last edition revised to date was published in 1916 (Series No. 269). The map is clearly printed and gives much accurate information. It covers also European Russia. This map is a great improvement on the larger scale map of communications.

The best small-scale map of Siberia is in two sheets in the *Atlas Marksa,* Petrograd, 1910. It shows Siberia on a scale of 1 : 10,000,000.

The Ministry of Agriculture, Emigration Department, published in 1914 a large *Atlas of Asiatic Russia,* with three volumes of text. It contains a number of economic maps but no new topographical work and no maps on a scale as large as 40 versts to an inch.

Special maps of parts of Siberia are noticed under the chapters to which they refer.

Of European Russia there are maps on a scale of 10 versts to an inch (1 : 420,000), of which a new edition, but with few corrections, was issued in 1914.

The Ministry of Ways and Communications publishes a *Map of the Railways, Roads, and Waterways of European Russia* on a scale of 40 versts to an inch. The last edition is dated 1913, and there are nine sheets, of which Nos. 2 and 3 cover most of Arctic Russia. This map is much better executed than the smaller-scale one of Asiatic Russia.

All the maps referred to above are in Russian. The only good map of Siberia in Roman characters is a small scale one (1 : 7,500,000) in *Stieler's Hand Atlas* (1916). It should be noted that the transliteration of Russian characters by the German system is liable to disguise many of the names.

Maps of the Amur River and Lake Baikal on a scale of 1 : 1,750,000 are included in Volume III.

CHAPTER I

GENERAL GEOGRAPHICAL FEATURES

Position, boundaries, and extent—West and East Siberia—Main geographical divisions—Rivers—Lakes—Coasts.

POSITION, BOUNDARIES, AND EXTENT

SIBERIA is bounded by the Ural Mountains on the west, by the Arctic and Pacific Oceans on the north and east. In the south-west the generally accepted frontier runs from the sources of the River Ural in the west across the Central Asian steppe lands to the Tarbagatai Mountains, and thence by a devious course that does not coincide with the watershed eastward to the River Argun and along the Argun and the Amur to the Pacific. The western part of this boundary, against Russian Central Asia, is an arbitrary one with no counterpart in geographical features. Siberia in fact is often taken to include the two eastern steppe provinces of Akmolinsk and Semipalatinsk. The northern but not the southern parts of these are certainly Siberian in character, while the same applies to the western steppe province of Turgai, which, however, is always excluded from Siberia. For the purpose of this book the steppes in general are included without adherence to administrative boundaries. The eastern part of the southern boundary is against Chinese Mongolia and Manchuria and near the Pacific for a few miles against Korea. In its western part the Uryankhai region south of the frontier is nominally Mongolian but actually in Russian occupation. The total land frontier is about 10,000 miles long and the sea frontier twice that length.

Siberia, excluding the steppe regions, has an area of about 4,800,000 square miles, which is $1\frac{1}{2}$ times the area of Europe, $2\frac{1}{3}$ times the area of European Russia, and 40 times the size of the British Isles. Its latitudinal limits are 49° N. to 77° 42′ N., and it stretches from long. 59° E. to 174° 24′ E. The Steppe regions included with Siberia in this book add about 450,000 square miles.

The name Siberia is supposed to be derived from the Russian word Сибиръ, which in the sixteenth century indicated the chief Tartar settlement on the Irtish, and was afterwards extended to include all Russian possessions in Asia. Later it was restricted to its present application.

West and East Siberia

Siberia may be conveniently divided into two unequal parts, western and eastern, of which western Siberia is the basin of the Ob, and though the more important of the two divisions, is only one-third of the area of eastern Siberia. The contrast between the south of Siberia with its great fertility, and the north with its barrenness almost as extreme, is easily recognized, but it is a contrast that holds chiefly in the west. The contrast between western and eastern Siberia is not so strongly marked and is often overlooked. The conception of Siberia as a vast plain rising with the gentlest gradient from the sea is true only for the west, or more strictly speaking for the Ob basin, and, in Arctic Russia, for the Pechora basin. East of the Yenisei these conditions do not hold. The elevations become considerable, and east of the Lena the surface is too irregular to be described as a plain. Low shores comparable with those in the west only occur about the mouths of the great rivers. In the extreme east the interior highlands reach the sea and leave only small and disconnected areas of plain along the coast.

Western Siberia extends through a great range of latitude and merges into the steppes of Central Asia and the plains of European Russia. Eastern Siberia is much narrower from north to south, and narrows progressively towards the east. Mountains cut it off from Central Asia and restrict its intercourse with the west. In the Amur basin it opens naturally to Manchuria.

Western Siberia has thus a large area in temperate latitudes not far removed from Europe, and fit for agriculture, while eastern Siberia lies mainly in more northern latitudes far removed from Europe, and its wide expanses of forests leave little scope for agricultural development. The physical link between western and eastern Siberia is Lake Baikal and the land route that with difficulty rounds its southern end.

MAIN GEOGRAPHICAL DIVISIONS

The main features of the relief of Siberia are comparatively simple, and consist of certain well-defined regions. Two high plateaux occupy the heart of Asia, and extending from extreme west to extreme east cover nearly two-fifths of the area of the continent. The western plateau, including Anatolia, Armenia, and Iran, is outside the region under consideration, but much of the eastern plateau, extending from the Pamirs and the Himalayas north-eastward towards the Bering Strait, lies within Siberia. This plateau includes Tibet, Chinese Turkestan, the Gobi desert, Mongolia, and much of Manchuria. A great part of it is desert and little of it is suitable for agriculture and human settlement. It forms a buffer land between that Asia which turns towards the Pacific and the Indian Oceans, tropical and sub-tropical in the main, and that which faces the Arctic Ocean and has its principal relations with Europe. The plateau lands have always formed an obstacle to the intermingling of the peoples on the two sides and have discouraged the meeting of east and west.

North of these plateaux and their high bordering ranges lies a broad alpine zone of rugged mountains and deep valleys. This zone averages about 150 miles in width, and its peaks rise to 5,000 or 6,000 ft. It includes the Altai Mountains and the Barguzinsk and other mountains around Lake Baikal.

North-west of the alpine zone comes the belt of high plains at an elevation of 1,500–2,000 ft. and with an average width of 200 miles, and beyond them, farther to the north-west, lie the low plains, not over 500 ft., which slope down gently to the Arctic Ocean.

Similar general physical features are repeated on a smaller scale to the south-east of the high plateaux.

Volcanic activity has played little part in Siberia. On the north-west border range of the high plateaux a few volcanic formations occur, but there are neither active volcanoes nor is there any historic record of one. In Kamchatka, however, the Pacific ring of volcanoes touches Siberia, and several active volcanoes occur, including Klyuchevskaya (16,130 ft.), said to be the loftiest volcano in Asia.

To these distinctive orographical features of Siberia may be added certain details.

The High Plateaux

The eastern plateau of Asia, the only one of the two which concerns Siberia, covers over one-fifth of the continent and extends 5,000 miles from south-west to north-east. It is widest in the west and middle, narrows towards the north-east, and is bordered by lofty ranges on all sides. Though called a plateau it is by no means of a uniform altitude, but is cut into terraces sharply defined from one another by escarpments which form ranges rising 500 to 1,000 ft. above the general level. The highest of these terraces is in the south and includes Tibet. It averages 12,000 to 13,000 ft. in height. Next in height is the terrace that lies on the north of the plateau, stretching from about long. 87° E. to long. 127° E., and includes north-west Mongolia and much of the Transbaikal region of Siberia with the Selenga, Vitim, and Aldan plateaux. This terrace has an average height of 3,000 to 5,000 ft. In addition to the escarpments which fringe the terraces there are a number of disjointed ranges, many of which run north-west and south-east, and others more or less parallel to the greater ranges. These still further diversify the surface of the plateau and give it the appearance of a region of great structural complexity. The plateau forms the water-parting between the Arctic and Pacific drainage. No stream crosses it from one side to the other. But on its surface there are few well-defined watersheds between the rivers except the escarpments which fringe the terraces. In many cases adjacent rivers are separated from one another only by marshes. In the east and north-east the high plateau is forested, but in the centre and the west it is a desert, and it is little more productive in the south.

The high plateaux of Siberia are built of gneisses, schists, clay-slates, and old limestones, all of Archaean and Palaeozoic age. On these old rocks occur in places Jurassic and Tertiary beds which are due to fresh-water lakes in those periods, at a time when the lowlands to the north were submerged and the proximity of the sea caused greater precipitation on the plateaux than is now the case.

The Great Border Ranges

Along the north-western and the south-eastern sides of the great plateau are continuous lofty border ranges. The north-western range is the most continuous in Siberia and in it occur the greatest heights. The continuity of this range is badly shown on most maps of Siberia, and its nomenclature is somewhat confused. The Tienshan, the Sayansk, the Ulan-Burgasi, the South Muya, and the Aldan Mountains are all parts of the north-western border range. It is 17–25 miles wide and 6,000 to 8,000 ft. high in the west, decreasing in the north-east to 4,000 and 3,000 ft. The south-east border range of the great plateau is known as the Great Khingan Range from China to the Amur River and thence to the north-east as the Stanovoi Mountains, including the Dzhugdzhur, Kolimsk,[1] and Anadir Mountains. There is still some doubt about the exact course of the northern end of the Great Khingan and its junction with the Stanovoi Range. In most maps of Siberia it is represented as crossing the Amur at the confluence of the Shilka and there terminating, while the Yablonoi Mountains are represented as running eastward from the head streams of the Olekma, forming the northern boundary of the Amur basin

[1] This is the original use of the name Kolimsk Mountains, which is more generally but erroneously applied to the range east of the Kolima River.

and joining the Stanovoi Mountains which continue to the north-east. There seems to be no information to support this suggestion, which originated arbitrarily in the days when little was known of the Amur basin. The supposed continuity of the Stanovoi and Yablonoi Mountains has no existence, and the Stanovoi Mountains in reality are of the same structure and origin as the Great Khingan Range, which crosses the Amur in the vicinity of the River Kumara, about 600 miles east of the confluence of the Shilka. More light, however, is needed on the junction of these ranges.

The Great Khingan Range is about 1,000 to 2,500 ft. above the level of the plains to the east, but it rises little above the general level of the plateau, so that, viewed from the west from the surface of the plateau, it has hardly the appearance of a mountain-range. Its crest is 3,000 to 4,500 ft. The Stanovoi Mountains are little known, and their north-eastern termination is uncertain. They seem to maintain the general features of the Great Khingan Range.

The continuity of the north-western range is broken in places by great trenches or gently graded slopes which give access from the plateau to the plains beyond. These trenches are the most important orographical features in Central Asia, for they link the lowlands of Siberia and the Transcaspian steppes with the high plateau and China. The most striking is the so-called Dzungarian trench, down which runs the head-stream of the Irtish from its source on the plateau to Lake Zaisan. As it descends the Irtish receives many tributaries from the Mongolian Altai, which stand above the northern side of the trench. The Dzungarian trench presents an easy route, and was one of the ways by which the Mongols spread west-wards to nearer Asia and Europe. Farther east the Selenga River descends to Lake Baikal in another trench and affords a much-used route via Kyakhta between Siberia and China. By the lower part of this trench the Siberian Railway climbs to the plateau and by an eastern branch of this trench, down which flows the Uda, the Siberian road ascends to Chita on its way to the east.

The Vitim, Olekma and Konam trenches are others along the range, named from the streams which flow down them.

The Alpine Zone

The broad zone of alpine highlands lying north-west of the plateau is a complex mass of ranges and spurs separated by deep valleys which are often swampy and strewn with boulders. The Altai, Baikal and Barguzinsk Mountains are part of this region. It has an average width of about 150 miles and a length of about 2,000 miles. The summits range from 5,000 to 8,000 ft. Towards the north-east it becomes lower and less rugged as it meets the Arctic Ocean. The so-called Kolimsk Mountains (see above) are the most striking features of the alpine foreland in the north-east. A longitudinal valley 10 to 25 miles wide is often noticeable between the border range and the alpine foreland. Most parts of this valley are occupied by secondary tributaries of the main rivers. The valley floor has an elevation of 1,000 to 2,000 ft. The alpine foreland is built of granites, syenites and crystalline slates. In it occurs the deep depression in which lies Lake Baikal. The greater part of the alpine foreland is densely forested, except where the peaks rise above the limit of the tree growth, and little of it is accurately known. The wild inhospitable nature of the region and the virgin forests offer few inducements to its penetration by man except where rich gold deposits occur, as in the Altai Mountains and the Yeniseisk district. In places, however, the fertility of the mountain valleys is attracting Russian colonists. Routes through the region are difficult. Much skill was required to carry the Siberian Railway round the south of Lake Baikal.

The alpine foreland occurs also to the south-east of the Great Khingan Range. It is 70 to 150 miles wide in China, but disappears towards the north-east and is lost below the waters of the Sea of Okhotsk.

The High Plains

Beyond the alpine foreland lies a broad zone, about 200 miles wide, of high plains at an elevation of 1,500 to 2,000 ft. They have few mountains, the only important ones being the Yeniseisk and Verkhoyansk Mountains, but the deep-cut gorges of the rivers draining from the plateau and the alpine zone give a hilly appearance to the plains. The plains are composed of more or less horizontal strata of Upper Devonian, Secondary and Tertiary ages in which the rivers have easily cut valleys 400 to 800 ft. deep. The high plains are forested in the wetter north-east but are steppe lands in the drier south-west.

On the south-eastern side of the plateau there is a belt of high plains about 100 miles wide. A range of granitic and schistose mountains called indifferently the Little Khingan, the Bureya and the Dousse Alin, runs along these plains parallel to the Great Khingan Mountains. Beyond these mountains, still farther east are the Sikhota Alin Range of the Maritime Province and the ranges crossing Sakhalin and Kamchatka.

The Low Plains

That part of Siberia which has most economic importance excluding only the Amur valley, and including practically the whole of western Siberia, comprises the low plains. They are seldom over 500 ft. in elevation, except where a few mountain ranges occur, and they slope gradually down to the Arctic Ocean. In some places an escarpment separates the high from the low plains: in other places the transition is gradual. The gradient of the plains is very gentle, being less than a foot per mile in the west. This gentle gradient is continued beneath the sea and gives shoal water far to the north of Siberia in the Arctic Ocean. Between the Urals and the Yenisei there are no hills of any importance. East of the Yenisei are the Pitski Range and the Tunguska Mountains on the borders of the high plains, the Syeverma Mountains (3,000 ft.) north of the Lower Tunguska, the Birranga Mountains in the Taimir Peninsula, the Vilyuisk Mountains west of the Lena, and the Verkhoyansk and Orulgan Mountains (4,000 ft.) east of the Lena. The highest peak in the Verkhoyansk Mountains is said to be 7,900 ft. Farther to the east the plains are more diversified and much narrower. They scarcely merit the name of plains as they merge into the plateau region of the north-east.

The low plains of Siberia are of Palaeozoic rocks deeply overlaid with post-glacial deposits showing that their emergence from the waters of the Arctic Ocean is recent from a geological standpoint. Many gigantic boulders scattered over their surface were no doubt dropped from floating icebergs.

The south-western part of the plains is semi-barren steppe-land beyond the confines of Siberia, but farther north these give place to rich meadow lands where the rainfall and black earth afford the best possible conditions for corn-growing and cattle-raising. Farther north the grass lands give way to forests, which in their turn thin out and disappear in the treeless swampy tundra which fringes the Arctic Ocean from Scandinavia to the Bering Strait. The tundra is useless for cultivation and settlement.

Across the western plains there are no obstacles to communication in any direction except the swamps of the tundra, which are impassable in summer, and the dense forests.

The Plains of Arctic Russia

The characteristic features of the low plains of Siberia occur farther west in Arctic Russia, but their continuity is interrupted by the low folds of the Ural Mountains which extend from about lat. 50° N to the Arctic Ocean. Their summits vary from 2,600 to 5,000 ft., and the greatest height is Telposiz (5,530 ft.) in about lat. 64° N. The Urals are low and wide, and consist of three discontinuous ranges in the south, between which there are many routes. North of lat. 60° N. they are more compact and continuous. The range becomes lower as it approaches Baidaratskaya Bay, and reappears in the low hills of the Yamal Peninsula. Two branches run from the Urals towards the north-west. The first is the Timan Range, which crosses the plains from about lat. 60° to 62° N. to the Kanin Peninsula. Its greatest height is about 750 ft., and its average height considerably less. The second range is the Pai-Khoi, which runs from about lat. 67° to 68° N. to Vaigach and Novaya Zemlya.

For the rest Arctic Russia east of the White Sea is low and swampy like north-western Siberia, and covered with forest and tundra.

The Kola Peninsula

West of the White Sea the country differs. The Kola Peninsula is geographically an extension of the mountainous region of Scandinavia. The interior is elevated and the north coast high and steep. Tundra in the north and meagre orests in the south cover the peninsula. South of the Kola Peninsula is the low-lying lake-studded region on the west of the White Sea, in reality an eastern extension of the Finnish lake plateau.

Rivers

Across the plains of Arctic Russia and Siberia many great rivers drain from the highland regions in the south to the Arctic Ocean. The uniformity in the direction of flow and the other characteristics of these rivers find their explanation in the relief of the land. The largest rivers are the Ob, Yenisei, and Lena with their many tributaries. They all rise in the central high plateau and drain through the alpine foreland to the plains across which they flow with sluggish, winding courses whose length is dependent on the breadth of the plains. Further east, where the highland region trends northward towards the coast, the rivers are necessarily shorter, swifter, and more direct; but the Yana, Indigirka, Kolima, and Omolon show on a smaller scale most of the characteristics of the rivers of the west. The narrowness of north-eastern Siberia and the proximity of the highlands to the sea cause the eastward drainage to flow in short rapid streams. The only exception is the Amur, which is comparable to the northern rivers. Like them it drains from the high plateau, across the alpine foreland and the plains. The chief respect in which it differs from the other great Siberian rivers, in addition to its Pacific outlet, is that a great part of the courses of the main river and the tributaries are on the high plains and the plateau.

The rivers of Arctic Russia, rising in the Urals, are necessarily shorter than those of western Siberia, but in other respects the Northern Dvina and the Pechora are similar to the Ob.

Importance of the Rivers

Their long courses over gently sloping plains give the Siberian rivers certain characteristics which have had a great influence on the history and development of Siberia. In the first place, the absence of a very decided slope means that the rivers wind a great deal, and have ill-defined watersheds which are easily crossed. In the second place, the gentle gradient of the plains makes the rivers slow and navigable almost to their sources. Lastly, the northward course of most of the rivers results in their waters swinging to the east, owing to the rotation of the earth, and as the rivers erode easily in the soft plain, their right or eastern banks are generally high and suitable for settlements, while their left or western banks are low, ill-defined, and liable to inundation. In their plain courses the rivers are sometimes several miles wide, shallow, and studded with sandbanks and islands, which are often completely inundated in times of flood. Their channels change from year to year, and their depth varies with the season. Yet despite all drawbacks the rivers form the chief highways of Siberia, and their value is enhanced by the vastness of the plains, the dense forests, and the swampy tundra, all of which make land travelling difficult if not impossible. There are no towns of any importance in Arctic Russia and Siberia which are not on navigable waterways. Of all the physical features of Siberia, it is the rivers that have had the most progressive influence on the country, the Ob, the Irtish, and the Yenisei most, and the Lena least of all the great rivers; but as population spreads eastward even the Lena will take its share. Railways will never replace waterways: they will make east and west routes, but will feed and be fed by the waterways. Railways link Siberia to the outer world, and in a measure advertise it, but the rivers do the work of real progress and settlement.

The chief disadvantage of the Siberian rivers is that most of them flow north, and so give access only to the Arctic Ocean, and all of them are closed to navigation by ice for more than half the year. The northern exits of the Ob, Yenisei, and other rivers have so far made them of use principally for internal commerce, and militated against their value as gateways of external trade. But there are signs that this will not always remain the case to the same extent (see Vol. II, Chapter IV).

In the southern part of the plains, the region of most value for human settlement, the tributaries of the different systems closely approach one another, and the basins are so dovetailed that by short and easy portages there is water communication from one end of Siberia to the other. This facilitated the penetration of Siberia by the Cossacks. Passing from the Ob to the Yenisei and Lena they reached the Sea of Okhotsk. And it was probably due to their missing the Amur and being led further north by the waterways to uninviting shores that Russia's entry into the Amur basin was so long delayed. Her claims on the Ussuri and the Yellow Sea coast came too late to be firmly established before Japan had begun to look westward.

The extent of the Siberian plains is so vast that many parts are at a considerable distance from a navigable waterway. In the north, where the rivers are fewer, and many run direct to the sea, this isolation of certain areas is most marked. Some of the northern regions away from rivers or drained by small independent streams are almost unknown except to wandering tribes.

Lakes

Siberia has many lakes, particularly in the west. The largest lie in the alpine foreland in the south. Lake Baikal covers an area of nearly 13,200 square miles, and is 400 miles long and 18 to 66 miles broad. Lake Zaisan, which lies outside Siberia proper, on the course of the Irtish, is 707 square miles in area. The Kirghiz and Baraba steppes are dotted with small lakes, many with ill-defined margins. On the Selenga and Vitim plateaux there are also many small lakes. In the lower part of the Amur basin are several larger lakes, including Lake Kada and Kizi near the mouth of the Amur, Lake Odzhal further up, and Lake Khanka (1,700 square miles) in the Ussuri valley, partly within Chinese Manchuria.

Coasts

The coast-line of Siberia has a great length, but little of it is important, as traffic to and from Siberia is principally overland via European Russia. The northern coast-line is still imperfectly charted except in the extreme west. It is blocked with ice for the greater part of the year, and in no month is navigation free from difficulties on account of ice. The coast of Russia west of the White Sea is the only part of the Russian Arctic coast which is approachable all the year round.

The Pacific coast of Siberia is less inhospitable, is faced with deeper water, and has several good harbours, but it opens to an unproductive hinterland, and is blocked with ice for much of the year. These drawbacks decrease progressively towards the south. Consequently the most important seaport, Vladivostok, lies at the extreme south of Russian Pacific territory.

The coasts of Siberia and the off-lying islands are described in detail in subsequent chapters in Volumes II and III.

CHAPTER II

CLIMATE

General characteristics—Temperature—Pressure and winds—Precipitation—Climate and Agriculture—Climatic Regions—Freezing and thawing of rivers of Siberia.

General Characteristics

The climate of Siberia is typically continental, and is characterized by a great range of temperature between winter and summer; a reversal of pressure conditions, and consequently of winds, between winter and summer; and a small amount of annual precipitation. In a general way it is comparable with the climate of European Russia, but is more extreme in all respects. The winter is long and very cold, but generally calm and dry with little cloud to interfere with the bright sunshine. The chief populated parts of Siberia lie between lat. 50° N. and lat. 60° N., and so receive, roughly speaking, as much insolation as the British Isles, but the extreme north has a certain period of darkness in midwinter. January is the coldest month. Snowfall is seldom deep.

The months of vegetative growth are May to October, with a mean temperature of about 56° F. in western Siberia, and about 53° to 60° F. in eastern Siberia, but only the three months of June, July, and August can really be regarded as summer. Most of the precipitation occurs in that period. In September the temperatures begin to fall rapidly. Spring and autumn are short seasons and are scarcely noticeable except in the south.

The causes of these characteristics are to be found in the physical circumstances of the country. It is a compact land mass open to the north by gently sloping plains, but closed to the south by mountains which serve to a great degree as climatic barriers. The moderating influence of the ocean can be felt only in the extreme east, where the Dzhugdzhur and Stanovoi Mountains lie near the coast, and prevent the oceanic influences penetrating far inland. The only ocean to which the plains of Siberia lie open is the cold Arctic Ocean, which is so encumbered by ice for a great part of the year, that it has little beneficial influence on the climate. Lastly, the country rises towards the south, except in the extreme west, and so the temperatures are lower than they would otherwise be, despite the comparatively low latitudes of that part of Siberia.

Temperature

The mean annual temperature of practically the whole of Siberia is below 36° F., and of all, except the extreme south, below 32° F., but these figures convey little because of the great seasonal range of temperature. The winter temperatures are much lower, and the summer temperatures slightly higher than the latitudes suggest. The greatest extremes occur in the north-east between the Aldan and the Arctic Ocean, where Verkhoyansk, in lat. 67° N., has a January mean of −60·7° F. and a July mean of 59·7° F., or a range of over 120° F., probably greater than occurs elsewhere on the face of the globe. Other places in eastern Siberia much farther south experience very low winter temperatures, such as Yakutsk in lat. 62° 1′ N., which has a January mean of −46·0° F. From this pole of cold in the Yana and Lena region winter temperatures increase in all directions. Even to the north along the shores of the Arctic Ocean the mid-winter temperatures are not so low as at Verkhoyansk. At Sagastir in the Lena delta, in lat. 73° 23′ N., the mean of February, the coldest month, in two years' observations, was −36·4° F., and the *Fram*, in her drift across the Arctic Ocean, had a January mean no lower than −31·9° F. In the east the waters of the Pacific in winter carry comparatively high coastal temperatures north as far as the Chukchee Peninsula, but the fall westward to the low temperatures of the interior is rapid. Thus the January mean of Vladivostok is 4·8° F., of Petropavlovsk in Kamchatka 13·8° F., but of Nikolaevsk −10·1° F. To the west and particularly the south-west of the pole of cold the increase in winter temperatures is more gradual. Yeniseisk has a January mean of −10·1° F., Tomsk of −3·3° F., and Tobolsk of −2·2° F. Even in the Steppe provinces the January mean ranges from −4° F. in the north to 17·5° F. at Lake Balkhash in about lat. 45° N. A January mean above freezing point does not occur north of Tashkent and Bokhara near the southern frontier of Asiatic Russia. In north-west Siberia the Atlantic influences make themselves felt to a small extent; thus Berezov on the Ob in about lat. 63° 40′ N. has a January mean of −10·6° F., and Arkhangel in 64° 32′ N. has a January mean of 7·5° F. This influence is considered further on p. 34. The freezing of Siberian rivers is considered at the end of this chapter, and ice in the polar seas in Vol. II, Chapters I and IV, and Vol. III, Chapter II.

In summer the highest temperatures occur in the south and south-west, and there is a decrease towards the north and east coasts, but the distribution of temperature conforms to latitude much more than in winter. The shores of the Arctic Ocean have a July and August mean generally well above freezing point, thus Sagastir in the Lena delta has a July mean of 40·3° F., and the *Fram*, in the Arctic Ocean, had a July mean no lower than 32° F. As far south as the Arctic Circle the increase of temperature is comparatively rapid, but south of the Arctic Circle it becomes more gradual. In the east temperatures as a rule are slightly higher than in corresponding latitudes in the west, but this small difference disappears in the south. The extreme south of Siberia has a July mean of over 71° F., and in the Steppe provinces the July mean goes as high as 80° F. On the east coasts the Pacific makes itself felt as a cooling influence, and the July isotherms, like the January ones, run roughly parallel with the coast from Sakhalin to the mouth of the Anadir. Thus Okhotsk in lat. 59° 21′ N. has a July mean of only 55·2° F., or about 11° F lower than Olekminsk, in the interior in much the same latitude, and Petropavlovsk in lat. 52° 53′ N. has an August mean of 58·3° compared with 66·9° F. at Yeniseisk on about the same parallel. Despite the low summer pressure over Siberia these maritime influences do not penetrate far inland. Lake Baikal exercises locally the functions of a sea in reducing summer temperatures in its vicinity.

Pressure and Winds

The low winter temperatures of Siberia result in an extensive high-pressure system developing over the country at that season. The frozen ocean to the north aids in its development. In January the highest pressure lies SW. of Lake Baikal, and extends thence to the NE. and SW. The pressure decreases towards the NW. of European Russia, where a comparatively low pressure area extends from the Atlantic over the Barents Sea, and towards the east, where a wide low pressure system lies over the North Pacific. Lake Baikal causes a local weakening of pressure in the heart of Siberia. As a result of these pressure conditions over Siberia the winter winds as a rule are light, generally from the SW. in the north, and from the E. and SE. in the south. But calms are characteristic of a Siberian winter, and consequently the intense cold is tolerable, and has comparatively little ill effect on vegetation and human activities. In March the centre of high pressure moves northwards to the Arctic Ocean, and by April the pressure over Siberia and the Pacific is almost equalized, while in May the rapidly rising temperature results in the formation of an extensive low pressure system over the country, which reaches its greatest development in July. The reversal in pressure conditions results in inflowing currents of air throughout Siberia. In western Siberia westerly and northerly winds prevail, in the Taimir region cool northerly winds, and in the north-east easterly winds. On the Ussuri and Maritime regions there is a more marked monsoon effect owing to the close proximity of the cool ocean to the comparatively warm land. Vladivostok has a prevalence of warm south-easterly winds during summer.

As a result of the low pressure the summer winds of Siberia are often strong, and gales occur at that season. The mountains of the south form a fairly effective barrier against southerly air currents, but föhn winds, warmed by their descent from high altitudes, not infrequently blow in the northern valleys of the Altai and Sayansk Mountains.

Precipitation

Precipitation throughout the whole of Siberia is slight and occurs chiefly in summer. It is least in the far north, where it is less than 8 ins. in the year, and it increases towards the south, reaching its maxima of 18 ins. or over in the south-west and the Altai region, and 20 ins. or more in the Amur region. In Kamchatka, where the monsoon is well marked, the total annual fall is 40 ins. or more. In the Steppe provinces in the far south-west the annual precipitation again decreases towards the Sea of Aral, where it is under 4 ins. In the Tienshan Mountains, however, between the steppes and Chinese territory, the annual amount rises to three or four times that figure.

As regards seasonal distribution, summer, as already stated, is the time of most precipitation. On an average 50–55 per cent. of the annual amount falls during June, July, and August. The daily fall is generally slight. The southern part of the Steppe provinces are again an exception, for they receive most of their scanty rainfall in winter. The only parts of Siberia proper that receive an appreciable amount of winter precipitation are the Vasuigan swamps and the Ishim steppes in the west, and Sakhalin and Kamchatka in the east. The Arctic coast, the Transbaikal, and the upper Amur regions are particularly dry in winter.

Cloudiness is a general accompaniment of the rainy season.

Drought is characteristic of many parts of Siberia, although the country is well supplied with great rivers. It must be remembered, however, that these rivers have their sources in the mountains of the southern frontier lands, where rainfall is more abundant than on the plains; and furthermore, that their stoppage by frost for more than half the year, and the slight loss they suffer from evaporation, except for a few weeks, are factors which combine to conserve their water-supply. Lastly, the frozen subsoil of the greater part of Siberia and the gentle gradients of the plains, especially in the west, make drainage slow, and give the country a wet appearance despite the small amount of precipitation that falls.

Climate and Agriculture

The influence exercised by the climate on most aspects of human activity in Siberia is noticed more particularly in the chapters on agriculture and communications, but attention may be drawn here to a few more general relations between climate and agriculture.

The high summer temperatures, if they were unaccompanied by cloudiness and rainfall, would be disastrous to agriculture. As it is the clouds temper the heat, and the rainfall is so evenly distributed throughout the summer months that agriculture receives the maximum benefit from it. However, a small diminution in the annual rainfall is most serious, as there is no margin to spare. In the Steppe provinces, where the summer rainfall is slight, and the summer temperatures very high, agriculture can be practised only along the rivers of the far south where irrigation is possible. In the Amur region the abundant summer rains favour agriculture while the monsoon region, including Kamchatka, has too much rain, in relation to its low summer temperatures, for agriculture to flourish.

The scarcity of snow in winter, throughout most parts of Siberia, allows the ground to freeze to great depths even in the south of the country. A permanently frozen subsoil extends north and east of a line drawn from the Kanin Peninsula, on the White Sea, east by Berezov on the Ob to Turukhansk on the Yenisei, thence south-east to Ilimsk and round the north and east of Lake Baikal, and west to the Uryankhai region: the lower Amur, Ussuri, and Maritime regions are excluded from this area. In summer the surface soil, in the area so defined, thaws to certain depths. Tree growth is not prevented, as the roots spread out laterally when they reach the frozen soil. In fact, some of the finest forests of Siberia are in this region. Provided a district has a sufficiently long and warm summer, the frozen soil actually assists cereal cultivation. The short roots of cereals do not reach the frozen subsoil, which on the other hand ensures a supply of water in the upper layers, and so saves the crops from disaster in case of drought. Of course, over the greater part of northern Siberia cereal cultivation is impracticable on account of the shortness of the summer and the waterlogged soil.

In western Siberia, with its greater winter snowfall and its higher winter temperatures, the soil does not remain permanently frozen. In other respects, however, the lower winter temperatures of eastern Siberia are not more unfavourable to agriculture than the higher temperatures of western Siberia, for both are too low to allow work on the land in winter. Frosts which occur as late as early summer and as early as August or September are most injurious to agriculture. In some agricultural regions July is the only month in which frost never occurs.

Climatic Regions

While practically the whole of Siberia experiences the same type of climate, the country can be divided into certain climatic regions. These regions have no clearly defined boundaries, and they merge imperceptibly into one another. The differences between their climates is in degree rather than in kind. The regions are as follows:

1. *The Arctic region* stretching from Lapland through Arctic Russia and Siberia to Bering Strait and extending southward to about lat. 64° N. in the west, and about lat. 67° N. in the east. Summer is very short and the temperature does not rise above 60° F. Winter is long and cold with a January mean of −10° F. to −40° F. except in the west. Spring and autumn scarcely occur. Rainfall in summer and

snowfall in winter are both slight. This region has neither so severe a winter climate, nor so warm a summer climate as east-central Siberia (Region 3).

The climate of the Kola Peninsula and the White Sea district forms a sub-region characterized chiefly by a milder winter than the rest of the region. This is due to the warm Atlantic drift, the influence of which is felt chiefly on the Murman coast but to a lesser extent in the White Sea and Kanin region, and rapidly disappears on the mainland farther east. The winter climate of Novaya Zemlya, particularly on the west side, feels its influence in comparatively high temperatures and in amount of precipitation, and Franz Josef Land may do so in exceptional years. The summer climate of this sub-region differs little from that of Arctic Siberia.

Temperature and Precipitation in Arctic Siberia

				Precipitation (inches)	
	Jan. mean	*July mean*	*Range*	*June–Aug.*	*Annual total*
	° F.	° F.	° F.		
Franz Josef Land	−11·5	36·1	47·6	?	?
Obdorsk	−16·4	56·5	72·9	4·7	11·1
Tolsti Nos	−28·8	51·8	80·6	2·0	12·0
Turukhansk	−18·7	59·5	78·2	7·5	14·0
Sagastir: Lena delta	−36·4 (Feb.)	40·3	76·7	?	?
Fram [1]	−31·9	32·0	63·9	?	?

Temperature and Precipitation in Arctic Russia

				Precipitation (inches)	
	Jan. mean	*July mean*	*Range*	*June–Aug.*	*Annual total*
	° F.	° F.	° F.		
Kola	11·8	54·8	43·0	4·0	8·0
Arkhangel	7·5	60·4 (Aug.)	52·9	6·0	15·5
Kem	12·4	58·3	45·9	?	15·0
Troitsko-Pechorskoe	−1·1	60·2	61·3	?	?
Karmakul: Novaya Zemlya	2·3 (Feb.)	43·2	40·9	1·7	12·5

2. *West-central Siberia* is the chief populated region of Siberia and includes the south part of the Tobolsk Province,

[1] The mean of the temperatures taken in the *Fram* from October 1893 to July 1896 during her drift in the ice from between lat. 77° 30′ N. and lat. 85° 55′ N.

most of the Tomsk Province, the south of the Yeniseisk and Irkutsk Provinces, and the northern parts of the Steppe provinces. Both winter and summer are warmer than in the Arctic region. The January temperature varies from 0° F. to −10° F., and the July temperature from 64° F. to 70° F. Early morning frosts may occur as late as June or as early as September. The mean annual rainfall is about 16 to 20 ins., of which more than half falls in the three summer months. On account of the high temperatures summer, like winter, is a period of clear weather despite the rainfall.

Temperature and Precipitation in West-Central Siberia

	Jan. mean	July mean	Range	*Precipitation (inches)* June–Aug.	Annual total
	° F.	° F.	° F.		
Bogoslovski	−2·2	62·6	64·8	8·0	16·5
Tobolsk	−2·2	66·4	68·6	9·5	18·0
Ishim	−3·8	66·0	69·8	9·0	17·5
Narim	−7·4	67·4	74·8	?	?
Tomsk	−3·3	65·6	68·9	10·0	20·0
Kurgan	−1·5	68·7	70·2	?	?
Yeniseisk	−10·1	66·9	77·0	6·5	17·0
Krasnoyarsk	−3·6	66·7	70·3	?	?
Barnaul	−2·2	67·1	69·3	5·0	12·0
Irkutsk	−7·4	65·1	72·5	8·5	16·0

3. *East-central Siberia* is the largest region and includes most of the Yakutsk Province except the extreme north, the north of the Irkutsk Province, and the Transbaikal Province. The climate is the most extreme in the whole of Siberia, and is characterized by the great severity of the winter rather than by exceptional warmth in summer. In January the mean temperature ranges from −60° F. in the north to −4° F. in the south, but some parts of the south on account of their considerable elevations have a January mean much lower than −4° F. The absolute minimum recorded is −90° F. at Verkhoyansk. Several months may occur during which the temperature remains below −20° F., but on the other hand great ranges may occur in any month except July. July has a mean temperature ranging from 60° F. to 70° F

Precipitation varies from an annual total of 5 or 6 ins. to 17 or 18 ins. Snowfall is nowhere heavy and the Transbaikal Province is almost snowless.

Temperature and Precipitation in East-Central Siberia

	Jan. mean	*July mean*	*Range*	*Precipitation (inches* June–Aug.	Annual total)
	° F.	° F	° F.		
Kyakhta	−18·4	67·1	85·5	7·5	10·0
Olekminsk	−33·3	66·0	99·3	7·5	11·0
Yakutsk	−46·0	66·2	112·2	6·5	12·0
Verkhoyansk	−60·7	59·7	120·4	?	5·0
Novi-Selenginsk	−15·7	70·5	86·2	?	?
Verkhne-Udinsk	−17·3	66·2	83·5	4·5	8·0

4. *Amur and South-east region.*—This has a somewhat anomalous climate, for not only is much of the region farther south than any other part of Siberia proper, but it is the only region that is influenced by the ocean to any great extent. The Amur valley shows climatic features intermediate between those of the Transbaikal and the south-east coast region, which has January means above zero and in which thaws may occur in any month. A few miles inland the continental low temperatures occur. Strong winds on the coast may make the winter, despite its higher temperatures, much more unpleasant than in the colder but calmer interior. Winter minima as low as −27° F. have been recorded at Vladivostok. July means are about 65° F. to 70° F., decreasing towards the north, but summer may be chilly on account of strong wet winds. Monsoon influences cause heavy summer rainfall, decreasing from the coast inland. As a rule more than half the total annual precipitation falls in summer. Dense fogs are common on the coast in summer.

Sakhalin and Kamchatka are extreme examples of this type of climate, but their sea-girt position redeems them from the severity of the continental winter and mitigates the summer heat. Rainfall is heavy all the year round. Ayan, on the west coast of the Sea of Okhotsk, with a total annual fall of 44½ ins. gets the same heavy rainfall, but the northern coasts of the Sea of Okhotsk get comparatively little.

Temperature and Precipitation in the Amur Region and on South-east Coast

	Jan. mean	July mean	Range	Precipitation (inches) June–Aug.	Annual total
	° F.	° F.	° F.		
Nerchinsk	−21·3	65·3	86·6	11·5	16·0
Blagovyeshchensk	−13·7	70·3	84·0	11·5	20·0
Khabarovsk	−13·2	69·4	82·6	12·5	22·0
Sofiisk	−30·8	59·5	90·3	?	?
Nikolaevsk	−10·1	62·2	72·3	6·5	17·5
Vladivostok	4·8	69·4	64·6	6·0	15·0
Olgi Bay	8·9	68·0	59·1	9·0	20·5
Alexandrovsk (Sakhalin)	− 0·6	62·0 (Aug.)	62·6	7·5	22·5
Okhotsk	−15·9	55·2 (Aug.)	71·1	3·5	8·0
Petropavlovsk (Kamchatka)	13·8 (Feb.)	58·3 (Aug.)	72·1	6·5	48·0

5. *The Steppe region* of the south-west has extreme continental conditions of climate but with great summer heat more marked than severe winter cold. The January means range from about zero to 10° F. and the July means from 70° F. to 80° F. Rainfall is slight at all seasons and much of the region is practically a desert. However, the best agricultural region in Siberia lies where the steppe merges into West-central Siberia (Region 2). Strong winds sometimes occur in winter with drifting snow and in summer with driven sand. Only the northern part of the Steppe region is considered in this book.

Temperature and Precipitation in the Steppe region

	Jan. mean	July mean	Range	Precipitation (inches) June–Aug.	Annual total
	° F.	° F.	° F.		
Semipalatinsk	0·5	72·0	71·5	2·5	7·5

Freezing and Thawing of the Rivers of Siberia

The rivers are frozen over in 5–20 days of frost, the length of time varying with the severity of the frost. It is noteworthy that the smaller polar rivers, especially those lying between the Ob and the Yenisei and the Yenisei and the Lena, freeze far more rapidly than these great rivers with their enormous basins and warmer waters coming from the south. Rivers like the Taz or the Khatanga are unable to attain a high temperature during the short summer. The early freezing of the rivers between Lake Baikal and the Pacific is probably the result of the mountainous character of these regions, where cold is felt earlier than in the neighbouring districts. Several small rivers and streams flowing into the Verkhne-Vitim in the marshy Bargunsk forest, some of them running through deep ravines, remain covered with ice throughout the year.

The tables here given for the opening and closing of the rivers do not correspond with the opening and closing of navigation. The rivers are not navigable for at least a week after the breaking up of the ice, and often for a fortnight or more before the river is actually frozen over.

The dates given are the average for a varying number of years. An estimated date is given in brackets where actual figures were unobtainable. The dates according to the Russian calendar would be 13 days earlier than those given here.

WESTERN SIBERIA

River.	*Average date of breaking up of ice.*	*Average date of freezing over.*
Atbasar at Atbasar	May 7	—
Ayaguz at Sergiopol	April 5	Nov. 27
Biya at Biisk	May 5	Oct. 23
Charish at Byeloglazovo	„ 4	„ 21
Chulim at Ust-Chulimskaya	„ 8	„ 23
„ „ Nazarovskoe	„ 12	„ 18
„ „ Bogotolskoe	„ 9	„ 19
„ „ Achinsk	„ 8	„ 23
„ „ Tutalskoe	„ 18	„ 14
„ „ Ziryanovskoe	„ 12	„ 19
Irbit at Irbit	„ 5	„ 11
Irtish at Lake Zaisan	April 30	„ 21
„ „ Krasnoyarsk	„ 25	Dec. 6
„ „ Ust-Kamenogorsk	„ 30	„ 2
„ „ Semipalatinsk	„ 30	Nov. 26
„ „ Yamishevskaya	May 2	„ 24
„ „ Pavlodar	„ 6	„ 22
„ „ Omsk	„ 14	„ 18
„ „ Tara	„ 15	„ 18
„ „ Tobolsk	„ 14	„ 20
„ „ Samarovskoe	„ 28	„ 19

River.	Average date of breaking up of ice.	Average date of freezing over.
Iset at Yekaterinburg	May 11	Nov. 12
Ishim at Akmolinsk	,, 5	,, 17
,, ,, Petropavlovsk	,, 11	,, 13
,, ,, Ishim	,, 13	,, 14
Kartisak at Kartisak	,, 3	,, 9
Kiya at Mariinsk	,, 12	,, 13
Ob at Barnaul	,, 8	,, 23
,, Kolivan	,, 14	,, 21
,, Kolpashevo	,, 15	(Nov. 21)
,, Narim	,, 22	Nov. 18
,, Timskoe	,, 20	(Nov. 18)
,, Aleksandrovskoe	,, 26	Nov. 16
,, Surgut	,, 30	,, 15
,, Pesk fishing station	,, 24	,, 16
,, Kondinskoe	,, 28	,, 16
,, Obdorsk	June 16	,, 9
Om at Kainsk	May 17	Nov. 14
,, Omsk	,, 14	,, 13
Pishma at Tahtsa	,, 4	,, 17
Polui at Obdorsk	June 11	Oct. 27
Pyazina at Vedenskoe	—	Nov. 1
,, ,, Zaostrovskoe	July 9 (one year)	Oct. 14
,, ,, mouth	Aug. 4 (one year)	,, 7
Sosva at Berezov	June 3	Nov. 12
Tavda at Nikolsk factory	May 15	,, 17
Taz	June 10	,, 4
Tobol at Svyerinogolovskaya	May 5	,, 21
,, ,, Kurgan	,, 8	,, 16
,, ,, Yalutorovsk	,, 7	,, 20
,, ,, Blinnikova	,, 7	,, 14
,, ,, Tobolsk	,, 20	,, 17
Tom at Kuznetsk	,, 10	,, 29
,, ,, Polomoshnaya	,, 9	,, 25
,, ,, Tomsk	,, 13	,, 17
Tura at Verkhoture	,, 13	,, 11
,, ,, Turinsk	,, 7	,, 17
,, ,, Tyumen	,, 8	,, 12
Ui at Troitsk	April 27	,, 20
Uvelka at Troitsk	May 1	,, 17

EASTERN SIBERIA

River.	Average date of breaking up of ice.	Average date of freezing over.
Abakan at Ust-Abakanskoe	April 30	Nov. 30
Aldan at Ust-Maiskaya	May 31	Oct. 30
Amga at Sulgachinskskaya	June 2	(Nov. 30)
,, ,, Amginskaya	May 29	Dec. 4
Amur at Pokrovskaya	,, 13	Nov. 19
,, ,, Albazin	,, 16	,, 18
,, ,, Chernyaeva	,, 16	,, 26

River.	*Average date of breaking up of ice.*	*Average date of freezing over.*
Amur at Blagovyeshchensk	May 12	Nov. 25
,, ,, Raddevka	,, 12	(Nov. 25)
,, ,, Yekaterino-Nikolskaya	,, 10	Nov. 28
,, ,, Mikhailo-Semenovskaya	,, 2	(Dec. 2)
,, ,, Khabarovsk	,, 6	Dec. 7
,, ,, Mariinsk	,, 23	Nov. 24
,, ,, Nikolaevsk	June 1	,, 25
Anadir at Markovo	,, 19	Oct. 27
Anabar at mouth of River Krilya-Kan	—	,, 20
Angara (Upper Tunguska).		
,, at Irkutsk	April 22	Jan. 25
,, ,, Usolskoe	May 11	,, 8
,, ,, Olonskoe	,, 11	,, 3
,, ,, Verkhne-Ostrovskaya	,, 10	,, 5
,, ,, Balagansk	,, 20	Dec. 20
,, ,, Malishevka	,, 19	,, 21
,, ,, Shiveri	,, 23	,, 17
,, ,, Ust-Udinskoe	(May 24)	,, 15
,, ,, Podvoloshnaya	May 27	,, 13
,, ,, Bratski-Ostrog	,, 24	,, 7
,, ,, Pyani Poroga	(May 27)	,, 6
,, ,, Padunskoe	June 1	,, 8
,, ,, Shmanek	May 26	,, 9
,, ,, Voroveva	(May 26)	,, 5
,, ,, Karapchanskoe	May 26	Nov. 29
,, ,, Kezhemskoe	,, 28	,, 15
,, ,, Boguchanskoe	,, 26	,, 18
,, ,, Pinchuga	,, 26	,, 22
,, ,, Ribinskoe	,, 29	(Nov. 30)
Argun at Olochinskoe	April 26 (1875)	Nov. 19
,, ,, Argunskaya	May 16	,, 21
,, ,, Urovskoe	,, 17	,, 20
,, ,, Pokrovskaya	,, 17	,, 16
Balei at Gorokovskoe	April 29	,, 12
Barguzin at Barguzin	May 10	,, 12
Biliktui at Biliktuiskoe (mouth)	,, 9	,, 29
Biryusa at Biryusa	,, 15	,, 21
,, ,, Kontorskoe	,, 13	,, 27
Byelaya at Maltinskoe	,, 11	,, 20
Chikoi at Baikhor	,, 15	,, 21
,, ,, Kudarinskaya	,, 8	,, 23
Chima at Cheremkhovskoe	,, 16	,, 16
Gizhiga at Gizhiga	June 7	,, 2
Ilga at Znamenskoe	—	,, 14
Ilim at Nizhne-Ilimsk	May 18	,, 9
Indigirka at Russkoe Uste	June 29	Oct. 16
Ingoda at Titovo	May 13	Nov. 15
,, ,, Chita	,, 10	,, 17
,, ,, Kaidalovo	,, 14	(Nov. 19)

River.	*Average date of breaking up of ice.*	*Average date of freezing over.*
Ingoda at Knyaze-Beregovaya	May 16	(Nov. 19)
,, ,, Raz-Makhnina	,, 17	Nov. 20
Irkut at Shchinkovskoe	,, 2	,, 18
,, ,, Tunkinskoe	,, 11	,, 18
,, ,, Smolenskoe	,, 14	,, 19
,, ,, Irkutsk	,, 14	,, 5
Iya at Tulunovskoe	,, 15	,, 23
Kan at Kansk	,, 11	,, 25
,, ,, Antsiferskoe	,, 12	,, 18
Khara-Ulakh at mouth	June 23 (one year)	—
Khatanga, near mouth	June 28 (one year)	Oct. 12
,, at Khatangskoe	July 7	,, 19
Khilok at Petrovski Zavod	May 23	Nov. 17
Khor	April 30	Dec. 13
Kirenga at Kirensk	May 21	Nov. 22
Kolima at Urocheva	June 6	Oct. 25
,, ,, Sredne-Kolimsk	,, 11	,, 24
,, ,, Nizhne-Kolimsk	,, 18	,, 17
Kukhtui at Okhotsk	,, 2	Nov. 29
Kuta at Ust-Kutskoe	May 20	—
Lena at Kachugskoe	,, 17	Nov. 17
,, ,, Verkholensk	,, 19	,, 17
,, ,, Ust-Ilginskaya	,, 20	,, 13
,, ,, Ust-Orlinskaya	,, 18	,, 9
,, ,, Omoloevskaya	,, 18	,, 14
,, ,, Kirensk	,, 24	,, 4
,, ,, Vitimsk	,, 25	,, 21
,, ,, Nyuiskaya	June 3	,, 19
,, ,, Olekminsk	,, 1	,, 19
,, ,, Yakutsk	,, 10	,, 12
,, ,, Markha	,, 6	,, 10
,, ,, Bulun	,, 16	,, 2
,, ,, (mouth)	July 8	Oct. 15
Lower Tunguska, *see* Angara.		
Maya at Ust-Maiskaya	May 29	Nov. 10
Nercha at Nerchinsk	,, 14	,, 9
Oka at Ziminskoe	,, 15	,, 23
,, ,, Bratski-Ostrog (mouth)	,, 23	,, 19
Olekma at Troitskoe	,, 30	,, 16
Olenek at mouth	July 16 (one year)	Oct. 13
Onon at Aksha	May 5	Nov. 25
,, ,, Ust-Ilya	,, 12	,, 20
Oya at Yermakovskoe	,, 1	,, 24
Penzhina at Penzhina	June 12 (one year)	Oct. 23
Pyasina at Vedenskoe	—	Nov. 1
,, ,, Zaostrovskoe	July 9 (one year)	Oct. 14
,, ,, mouth	Aug. 4 (one year)	,, 7
Selenga at Novi-Selenginsk	May 8	Dec. 1
,, ,, Verkhne-Udinsk	,, 9	Nov. 22

River.	*Average date of breaking up of ice.*	*Average date of freezing over.*
Shilka at Mitrofanova	May 10	Nov. 23
„ „ Monastirskoe	„ 17	„ 17
„ „ Stryetensk	„ 11	„ 21
„ „ Pokrovskaya	„ 18	„ 15
Suifun at Razdolnoe	April 18	Dec. 11
Suputinka at Nikolsk-Ussuriski	„ 16	„ 4
Taimir at mouth	July 23 (one year)	Sept. 29
Tuba at Kuraginskoe	May 7	Dec. 6
(Lower) Tunguska at Preobrazhenskoe	„ 21	Nov. 1
„ „ „ Monastirskoe (mouth)	June 10	Oct. 31
Turukhan at Yanov	„ 11	„ 13
„ „ River Bayukta mouth	„ 11	„ 21
„ „ Turukhansk	„ 6	Nov. 3
Uchur at Yarmank	May 24	—
Uda at Udski-Ostrog	„ 7	Nov. 25
Uda at Nizhne-Udinsk	„ 14	„ 17
„ „ Verkhne-Udinsk	„ 13	„ 25
Upper Tunguska, *see* Angara.		
Ussuri at Kozlovskaya	April 30	„ 28
„ „ Khabarovsk	May 6	Dec. 8
Vilyui at Nyurba	June 1	Oct. 29
„ „ Vilyuisk	„ 5	Nov. 2
Yana at Verkhoyansk	„ 11	Oct. 23
„ „ Kazache	„ 17	„ 9
Yenisei at Minusinsk	May 11	Nov. 29
„ „ Abakanskoe	„ 11	Dec. 4
„ „ Krasnoyarsk	„ 12	„ 5
„ „ Kazachinskoe	„ 15	Nov. 26
„ „ Yeniseisk	„ 19	Dec. 2
„ „ Nazimovo	„ 18	Nov. 28
„ „ Verkhne-Imbatskoe	„ 29	„ 21
„ „ Turukhansk	June 8	„ 11
„ „ Potapovskoe	„ 20	(Nov. 14)
„ „ Luzino	„ 19	Nov. 4
„ „ Selyakino	„ 19	(Nov. 14)
„ „ Tolsti Nos	„ 25	Nov. 3
„ „ Golchikha	July 2	„ 5
„ „ (mouth)	June 23	Oct. 30
Zalara at Zalarinskoe	May 4	Nov. 15
Zavitaya at Mikhailovskaya	April 25	„ 21
Zeya at Blagovyeshchensk	May 11	„ 26

CHAPTER III

VEGETATION

The Tundra—Coniferous Forests—The Amur Forests—The Pacific Forests—Alpine Vegetation—Wooded Steppes—Kirghiz Steppes—Trans-baikal Steppes.

The vegetation of Siberia remains in its original state, and man has effected few changes. Broadly speaking, there are three great types of vegetation, very unequal in the areas they cover. In the north along the Arctic Ocean is the tundra, while an immense forest covers the rest of Siberia, with the exception of the steppe-lands in the south-west. The first two of these divisions extend into Arctic Russia.

The Tundra

The tundra stretches in a band 20 to 200 miles wide from west to east of the Russian Empire along the Arctic shores. The southern limit averages about lat. 68° N., but in the Pechora and Ob basins the boundary recedes south to about the Arctic Circle, in the Khatanga basin it advances to about lat. 72° N., and in the far east it trends south to the northern end of Kamchatka. All the far north-east from about long. 160° E. to Bering Strait is covered with tundra. Tundra in an impoverished form occurs on the Arctic islands north of Russia and Siberia.

While the term tundra is often used to convey the sense of a treeless Arctic plain, it really has reference only to the special type of vegetation which is generally found associated with Arctic plains. Tundra frequently occurs also on considerable elevations, as in the Taimir region, the Chukchee Peninsula, or the far east generally.

The vegetation of the tundra is chiefly grasses, sedges, and herbaceous plants, among which are many bulbous species. Cold waterlogged soil and want of humus militate against plant life. The only trees are dwarf birches (*Betula nana*) and willows not exceeding a few inches in height and generally creeping on the ground. Low bushes of heath, azalea, and arbutus also occur, but only in sheltered nooks do the trees or bushes grow to any height. Mosses and lichens, including the reindeer 'moss' are numerous, and in the east are the principal plants. In general character the tundra is uniform from Russia to Bering Strait, and for that matter all round the Arctic Ocean, even if the species of plants differ. In level places it is more swampy than in the hilly parts, and peat forms extensively. It is frozen and snow-covered for 8 or 9 months in the year. In the summer the surface thaws, but the subsoil remains frozen: the tundra for a few weeks is bright with flowers and alive with insect life, including mosquitoes, but at that season it is almost impassable to man. It is quite useless for agriculture, and has no economic value except for reindeer breeding.

Coniferous Forests

The coniferous forests or *taiga* begin gradually where the tundra ends. The polar limit of trees is largely determined by dry cold winds in winter which are hostile to all tree growth. The forests never reach the north coast, but in some sheltered river valleys such as those of the Ob, Yenisei, and Lena, a few badly developed trees reach the delta, and, generally speaking, the rivers seem to carry the forests northward into the tundra region. The taiga in one form or another extends from the Pacific through Asia, Russia, Finland, and Scandinavia, to the Atlantic. Southward it extends to the confines of Siberia, where it gives way to the steppes of Mongolia and Russian Central Asia. The forests vary a good deal in appearance and species, but are everywhere with a few small exceptions either solely or principally coniferous.

In Arctic Russia the chief species are the Scots pine (*Pinus sylvestris*), the Norway spruce (*Abies excelsa*), and the silver fir (*A. pectinata*) ; in the Urals the Siberian fir (*Abies sibirica*), and the Siberian larch (*Larix sibirica*). The deciduous trees which occur are oak, elm, ash, maple, and apple.

In the drier parts of the taiga of western Siberia the commonest species are the Siberian fir, the so-called Siberian 'cedar', or stone pine (*Pinus cembra*), the spruce (*Picea obovata*), the silver fir, and the Siberian larch. These are the prevailing trees in the Yenisei basin, the basin of the upper Ob, and the Altai region. Deciduous trees are rare. In the wetter region of the basins of the middle and lower Ob and the lower Irtish the taiga is marshy and has thick, impenetrable undergrowth. Larch is rare, and the Siberian fir predominates, but there is also an admixture of deciduous trees such as birch and aspen. Thickets of poplar, alder, and willow fringe the streams, and there are some birches.

Berry bushes are frequent except in the swampy parts, and include the wortleberry, bilberry, Arctic bramble, raspberry, and red and black currants.

The Altai Mountains have the same coniferous forests as western Siberia, but they are more open and the trees grow tall. In places there is undergrowth, but as a general rule it is absent. The forests extend to an altitude of about 5,000 ft. on the southern and 6,000 ft. on the northern side. On the south the flora is richest, and rhododendrons and azaleas appear.

The southern part of the marshy taiga of the Ob basin is known as the Vasuigan swamps. They are most fully developed in the northern part of the region between the Ob and the Irtish, but also stretch north of the Ob, and consist of swamps covered with dense thickets of birch, alder, aspen, Siberian cedar, pines, and a few larches. The Russian name is *urman*. They are practically impassable except in winter. In time of spring floods these urmans are so much inundated that they are termed the Vasuigan Sea.

The forests of eastern Siberia are very uniform from the Yenisei basin eastward to the Amur region and the Stanovoi Mountains. In the north they merge gradually through a region of gnarled and stunted trees into the tundra. The Siberian fir and the eastern larch (*Larix daurica*) are the prevailing species, but the Siberian 'cedar' (*Pinus cembra*) and the Scots pine also occur. The spruce (*Picea obovata*) and the Norway spruce go as far east as the Lena. On the whole, however, the forest of east differs from that of west Siberia mainly in less luxuriant growth. Undergrowth is not so frequent, and the hilly nature of the country gives fewer areas of marshy taiga. Furthermore, the poor rainfall and the cold dry winds during the long severe winter do not favour tree growth. The forests on the whole are open and low. Along the river banks, however, and in more sheltered places the trees grow taller and thicker. In the upper Lena basin the forests are more of the western type, and the Scots pine and the Siberian larch grow to a large size.

The Amur Forests

East of Lake Baikal and in the Amur region the vegetation changes. The Transbaikal is a transition region between the eastern and western floras, but among forest trees all the widespread species of the northern taiga are found. In the eastern part foliage trees make their appearance. These are of species different from, though allied to, those which occur in the northern forests of Russia, the oak, the elm, and characteristic species of the hazel and wild apple. Bushes that are typical of Mongolia also make their appearance.

In the Amur region the divergence of the vegetation from that of eastern Siberia in general is more pronounced. The flora is characterized by a great variety of forms and by the luxuriance of some species. Plants which are typical of China and Japan occur, and even representatives of the North American flora. In the northern part of the basin the forests are like those of eastern Siberia in general, but different species appear, until nearer the Amur the forests have an entirely different appearance, owing largely to the presence of many deciduous trees. The eastern larch, the Siberian fir, the Siberian spruce, the Scots pine, and the yew occur as well as another species of spruce, the ayan pitch pine (*Picea ayanensis*), and the white cedar or Manchu pine (*Pinus mandshurica*), which replaces the *Pinus cembra* of the north and west. Among deciduous trees are the oak, elm, lime, maple, walnut, ash, aspen, willow, hornbeam, and apple, all of distinct eastern species, and the cork tree (*Phellodendron amurense*). In the upper and middle Amur regions deciduous woods are commoner than coniferous woods along the river, and wide meadows of natural grass land often occur, but in the lower Amur region coniferous forests again prevail.

The Sikhota Alin Range and the Maritime Province generally are also forested. In the north the forests are mainly of larch, Siberian fir, and white cedar, but in the south deciduous trees are more common, and the oak as well as the Siberian cedar are the characteristic species. These forests are typical of the Ussuri region.

The Pacific Forests

The northern taiga continues to the shores of the Okhotsk Sea, and occurs in Kamchatka and Sakhalin. The eastern larch (*Larix daurica*) is the prevailing species, but the Siberian fir and Siberian cedar are common, mingled with a few birch, alders, and shrubs, including the clematis, wild rose, and honeysuckle. The herbaceous vegetation is more prominent than farther west, and the forests undergo a change in appearance. The upper limit is at a low altitude, and the trees are dwarfed on account of the strong winds.

In the forests of Kamchatka the trees are more widely spaced, and the Siberian fir and the cedar are the commonest species, mixed with which are a few deciduous species including the birch, alder, and poplar. Some natural meadows occur.

Sakhalin, in the south, has forests like Kamchatka : in the north the forests rather resemble those of Okhotsk, the eastern larch being the chief species.

Alpine Vegetation

Alpine vegetation occurs at varying altitudes on the highest mountains in eastern Siberia. In the Altai it begins at about 6,500 ft.; in the Sayansk Mountains at about the same altitude, but in the Dzhugdzhur and Stanovoi Mountains and in the Verkhoyansk Mountains at gradually decreasing altitudes towards the north until it merges into the Arctic tundra. The general aspect of the alpine vegetation is much like that of the tundra except in the absence of swamps. Between the alpine vegetation and the forests lies a belt of rich mountain pasture, comparable with the high pastures of the European Alps and containing many of the same species.

Wooded Steppes

Steppe land is rare in Siberia proper, and occurs in large areas only in the west, where it is found to the south of the taiga. The Ural forests and the Vasuigan swamps give way to the Ishim and Baraba steppes, which extend in a strip about 100 to 200 miles wide from the Ural slopes to the Altai region. These are intermediate between the forests to the north and the true steppe lands to the south. Firs gradually disappear and are replaced by birches, aspens, and willows, which occur in clumps and along river banks, in a general expanse of rolling grass lands. In the northern part of the Ishim and Baraba steppes trees are frequent and swamps and urmans occur. Southward the swamps are replaced by numerous small lakes, many of which are saline, and trees become less frequent until in about lat. 55° N. true steppe lands begin. The Siberian Railway runs across the Baraba steppes. In the upper Ob basin these steppe lands are wider than to the west, and extend from the railway south to Semipalatinsk and east to Biisk and Kuznetsk.

The Ishim and Baraba steppes have a fertile soil consisting largely of black earth (*chernozem*), a kind of loess, rich in humus, and are the most valuable agricultural region in Siberia. A detached area of wooded steppes lies farther east between Achinsk and Minusinsk.

Kirghiz Steppes

True steppe lands begin south of the wooded steppe and lie outside Siberia proper. They cover the region known generally as the Kirghiz steppes. To the north, bordering the Baraba steppes, are the feather-grass steppes stretching from the Caspian steppe lands of Russia to the Yenisei basin, with a breadth of 200 to 270 miles. Trees are rare, and are represented only by a few birches, aspens, willows, alders, ashes, and poplars in the wetter places. Dwarf bushes, many of them thorny, are characteristic, including broom, hawthorn, and tamarisk. The herbaceous flora is rich and embraces many flowering plants. Feather grass (*Stipa*) is characteristic. To the south the climatic conditions peculiar to this region become intensified. Rainfall decreases and summer temperatures increase. The change is reflected in the vegetation, which assumes a semi-desert appearance as the steppes become more arid.

The steppe lands of western Siberia are chiefly of value for horse and cattle breeding. On account of the want of rainfall there is little scope for agriculture.

Transbaikal Steppes

In eastern Siberia natural steppe land is rare, for climatic conditions are more favourable to the development of forest, but in the Transbaikal region between the south of Lake Baikal and the River Argun there are considerable areas of steppe lands, in general appearance not unlike the mountain pastures of the Altai Mountains. This is best developed in the valleys. The higher ground of the Yablonoi and other ranges is covered with forests. These Transbaikal meadow lands are excellent for agriculture. In the south towards the frontier of Siberia they pass into the dry steppe lands of Mongolia.

CHAPTER IV

ANIMAL LIFE, FISHERIES AND HUNTING

Animal Life—Fisheries: A. Arctic Russian; B. Western Siberian; C. Eastern Siberian—Products of Wild Animals

Animal Life

THE land may be divided into four zones with special geographical characteristics, which are differentiated to some extent by their fauna. In the north (1) the Arctic or tundra zone, with ice or reindeer-moss, has its special group of animals; westward is (2) the taiga belt of coniferous trees, where in the depths of the forest there are not many wild animals; it is rather the skirts of the forest and the clearings made by the axe or fire to which animals resort, and in such clearings and at river fords the trapper finds them; (3) farther south the open steppe has fresh denizens, and finally (4) the high mountains of the Altai and Sayansk Ranges introduce a new element. No hard and fast line, however, can be drawn between these belts, and there is considerable overlapping.

Special Characteristics of Siberian Fauna

Among the characteristics of Siberian fauna are to be noted: (1) The animals as a rule are of greater size; some varieties are twice as large as the corresponding breeds in Europe. (2) White predominates, even among domestic animals: many animals, like the ermine, Arctic fox, and hare, are white during the winter months. (3) In exceptional circumstances there are huge migrations. Most famous are those of the lemmings, which move in vast armies regardless of natural obstacles, swimming estuaries, where they become the prey of killer whales or seals, and often plunging into the sea.

Land Mammals

There are found in the south the tiger, panther, snow-leopard (*Felis irbis*), lynx, and two varieties of wild cat. The tiger and snow-leopard do not reside north of Lake Balkhash, or the River Amur, but occasionally stray beyond. The tiger is not infrequently found near Chita and within 120 miles of Nikolaevsk, and is often met with in the jungles of Primorsk, in which 120 to 150 are shot or poisoned every year. Tigers gave great trouble during the construction of the Ussuri Railway. The natives view the tiger with superstition. The tiger is originally a Siberian and Mongolian animal, which has only reached India in comparatively recent times; the Indian tiger has much shorter hair than the northern variety. The lynx is comparatively common in all mountainous parts of Siberia; its fur is regarded as very valuable. The panther is found in the Primorsk forests. The steppe-cat is found in the Kirghiz steppes, and there are two varieties of wild cat in the Altai. It may be observed that the domestic Siberian cat is very fine, rivalling the Persian, but the European climate does not agree with it. In Tura black cats are specially bred for their fur.

The wolf is found all over Siberia; in the tundra it hunts the reindeer, and on the steppes the roe. The wolf is driven from the reindeer by shouting at it. Often a wooden clapper is hung round the neck of the leading reindeer: it makes a noise as though of a man chopping wood, and keeps the wolves at a distance till they have become habituated to the sound. Wolves, when hunting in small packs, are not dangerous to human beings. The solitary wolf is a retiring creature and anxious to avoid observation. In the north it is hunted not merely to protect the reindeer, but because the fur is valuable for caps, mittens, &c. The Alpine wolf is found in the Altai.

The fox is found in many species over the whole country. North of lat. 60° is the Arctic fox (*Vulpes lagopus*), distinguished by its short, blunt ears, long, bushy tail, and very long hair in winter. In the summer its upper parts are of a brownish colour; in the winter its whole coat turns white. It is found on the continent and on the islands of the Arctic Ocean, but is less common in eastern than in western Siberia. It is diminishing in numbers, owing to the reckless way in which it is trapped. The cubs are trapped in the burrow. All trapping of the fox-cub and sale of the cub-skins should be prohibited, nor should it be permitted to catch it when young, for its summer coat is of little value. The winter coat is exceptionally valuable for its downy fur and its colour. One variety of it is the blue fox, so called from its slaty colour, which is found chiefly in Arctic Russia and Kamchatka; it is valuable and scarce; its colour remains the same throughout the year, but its hair is longer in winter. The red fox of Siberia has a deeper, richer red than our variety, and a much more bushy tail. Its skin is esteemed, but not so much as that of the blue fox. Another very valuable species is the black or silver fox (*V. argentatus*) which has black hair with silver tips. The bulk of the skins exported by the Koryàks are fox skins. In the steppes, ranging from European Russia to the Amur, is the steppe-fox or *corsac*, a sort of representative of the Indian species. Foxes are usually killed with clubs or trapped, so as not to spoil the skin: they are also poisoned with strychnine.

The raccoon dog, a native of China and Japan, is found in the Amur basin: it is highly valued by the Manchu for its winter skin: in summer it is eaten.

Wild dogs are found in the south up to the snow-line.

Bears are found throughout Siberia. In the north is found the Polar bear (*Ursus maritimus*) wherever there are seals, which it hunts either along the beach or on the sea-ice. It arrives on floating ice, and lands on the coast, but does not penetrate far inland, and though known to have gone up the Gulf of Yenisei as far as Tolsti Nos, it generally does not leave the vicinity of sea-ice, for it is really a sea-mammal. Its fur is used for floor-rugs and is very durable. The brown or Kamchatkan bear (*Ursus beringianus*) is found across the centre of Siberia, wherever there are forests. It is hunted for fur and food. In the southern mountain ranges that adjoin the steppe district and by the Amur is found a black bear, (*Ursus tibetanus*) and several other species occur in the south. The bear is sometimes snared, sometimes hunted. Where fish abound, the huntsman waits for him by the rivers. The bear cage is a standing feature of the villages of many tribes, especially of the Goldi and the Gilyaks. The bear takes part in many religious ceremonies (see Chapter V).

The group of Mustelidae is of great economic value because of the trade in their furs. Most important of all is the sable (*Mustela zibellina*), a variety of marten. Its furs were the great lure into Siberia of the Russian trader in the sixteenth century, and for many years were the form in which *yassak* (tribute) was paid. It dislikes the proximity of human habitations and retires more and more to uninhabited parts, and has been seriously diminished in numbers. There are, however, many in Kamchatka. An order was made by the Russian Government that from February 1, 1913, to October15, 1916, no sables should be killed nor sable fur sold throughout Siberia. It is a difficult animal to hunt. It possesses a down which is entirely dark and of bluish tinge, and long, soft, glossy, black hair : the finest sable fur is tipped with silver. The farther north the sable is found the better is its fur, and it varies somewhat in tint in different parts of Siberia. It is found up to lat. 68° N. The best sables are found in the Vitimsk and Olekma regions, and in the neighbourhood of Nerchinsk and the headwaters of the Amgun and Zeya. The Sakhalin sables are of little value. White sable found in the Barguzin region is very rare and valuable. It is hunted in the beginning of winter. The kolonok (*Mustela sibirica*) is used as a substitute for the sable, especially to provide artists' 'sable' brushes. At Irbit fair 50,000 skins are sold annually. Other animals of the same type are the marten, of which there are three varieties in the Altai and upper Yenisei, and the cognate, but smaller weasel and polecat. The ermine is valuable, and is becoming correspondingly rare in some districts: the best ermines come from Ishim and the Baraba steppe: the skins are rarely over a foot in length, and they are sold in lots of 40 together, known as 'timbers'. The ermine is usually trapped. The glutton (or wolverine) is a much larger member of the same sub-family. It has powerful teeth, is almost entirely nocturnal in its habits, and is usually active through the winter: it swims streams and ascends rough-barked trees in search of food. It inhabits a belt across the middle of Siberia excluding the extreme north. It is also found in the north of Sakhalin. The glutton is no longer common. The skunk is taken for his fur in large numbers in the south of the Tomsk Government and in the Amur region.

The badger is found right across Siberia as far north as lat. 53° N.

The otter is found throughout Siberia up to the Arctic Circle and on Sakhalin, but is becoming rare. It is most hunted on the Amur for trade with the Manchu and Chinese, who value its fur highly.

The reindeer is most widely spread, being found not only in the northern tundras, but among the Soyots in the Sayansk Mountains, where there is plenty of lichen. The domesticated reindeer has been described elsewhere (see chap. V, p. 103). The wild reindeer is hunted for food in winter; it is also used for interbreeding with the domesticated animal.

In the north is also found the elk, the largest member of the family: the flesh is edible, the taste resembling something between venison and mutton. Its skin is also valuable; in many years there are 10,000 elk skins in the market of Yeniseisk. Its importance along the Amur is shown by its name *bayu*, a Tungus word meaning 'the animal'.

There are several varieties of the American wapiti (*Cervus canadensis*) in the south of the country. The number of wapiti has been greatly diminished by the sale of their horns to the Chinese, and the Russian Government have issued a prohibition which forbids their being shot on the Russian side of the frontier. The wapiti is distinguished by the great size of the fourth tine of the antlers.

Closely akin to the wapiti is the maral deer, which is domesticated in the Kirghiz country, the Altai, the middle Amur, and the Ussuri region : it is bred for the sake of its horns, which are sawn off when in the velvet, to produce from them a powder called *panty*, in high request among the Chinese for medicinal purposes. Western physicians believe that it has no medicinal value, but merely quickens the heart's action. It is sold for 30*s*. a lb. Another member of the family that produces an article of value is the musk-deer (*Moschus moschiferus*), which is hunted in the Altai, Sayansk, and Amur regions, and in Sakhalin, for the sake of the musk obtained from it. It is caught in winter, when the more vigorous climate makes it migrate from its ordinary home to something more accessible to hunters. Its flesh is eaten, its skin used for clothing, and its thin leg bones made into arrow-heads.

The roebuck is represented by two species, *Capreolus pygargus* in the upper Yenisei and Altai region, and *C. manchuricus*, which migrates from Manchuria into the Amur region in spring. It supplies abundant food. The skin and horns are sold to the Chinese.

Other members of the family found in Siberia are the common stag, rock-deer, spotted deer, and siku. The last named is found in the island of Askold near Vladivostock. There are several kinds of big-horn, but the *Ovis poli* of the Pamirs, which is sometimes said to be found in the Tienshan, is never really found so far north. The real *O. ammon*, or argali, is found in the Altai ; its wrinkled horns curl so much that they often make more than a complete circle. It lives amid the forest on mountains between 3,000 and 4,000 ft. high, and is difficult to secure. Its chief enemy, the wolf, hunts it to the neighbourhood of its lair, so that the young may have a share. Travellers who refer to argali among the lower heights of the Kirghiz region probably mean *O. sairensis* or some closely allied species. Between the Lena and Indigirka *O. borealis* occurs, and in Kamchatka *O. nivicola*. The saiga antelope (*Saiga tatarica*) extends as far east as long. 92° E., and another species is found in the Primorsk region. A gazelle

(*Gazella gutturosa*) is found in the steppes near Kosh-Agach: it is usually in colour of a light fawn, with white limbs, cheeks, underparts, sides, and rump; its tail is short, with a brown tip, and it has no dark face-markings, like most gazelles. Among the Altai there are a large but rapidly diminishing number of ibex (*Capra sibirica*), gradually being driven into the more remote valleys. It has very long horns, which are sold to the Chinese for the same purpose as those of the maral. The Kalmuks are very wasteful in hunting, and unless checked will exterminate the game of their region There are wild oxen and wild goats among the Sayansk Mountains.

The musk ox (*Ovibos moscatus*) is rare, if not extinct, in Siberia. It has been reported from Sakhalin.

The prickly hog is found in the southern portion of the steppe region and the steppe lands of the east. The wild boar occurs in the east mostly in Transbaikal and the Amur region.

The wild ass (*kulan*) is found upon the Kirghiz steppe: farther south, but outside the limits of this book, is found the famous Prezhevalski horse.

There are numerous species of rodents. Most important from the commercial point of view are the squirrels, the sale of whose skins is enormous: only the skins of Russian and Siberian squirrels are marketable. They are used for a great variety of articles, such as gloves, hoods, and carriage-aprons. The squirrels are of various colours, of which the black are most esteemed. Black squirrels eat mushrooms, brown squirrels cedar-cones, red squirrels hazel-nuts. Squirrels are found throughout south Siberia, especially in the forests, but there are none in Kamchatka. Besides the common squirrel there are to be found striped squirrels and flying squirrels. In the centre of Siberia the squirrel is much hunted. In the Transbaikal Province three million are killed annually. Squirrels are usually shot with guns having a small bore, and with bullets the size of a pea, so as to injure the skin as little as possible. The marmot found in the south, though much more frequently beyond the frontier, is of commercial value, because of the export of its skin to Europe, where it is sold as imitation sable. A variety of the marmot is the *tarabagan*, which is common in Transbaikal. Related to the marmot is the bobac (*Arctomys bobac*) which lives in the plains and stony hill lands, and the suslik, alike the friend and enemy of man, the former by reason of its flesh, which is esteemed a delicacy by the dwellers in the steppes, and the latter because its energetic burrowings make it a pest to agriculturists. It has been found specially injurious along the Lena. It is sometimes called the pouched marmot, because of its big cheek-pouches. Both it and the lemming (*Lemmus obensis*) store their food in winter. The lemming is said to protect its food against the depredations of other animals by covering it with poisonous plants. In eastern Siberia the Kamchadals remove the creature's store of grain and roots, but replace it with caviar or remains of fish, so that a regular trade is instituted between man and beast. The migrations of the lemming have been already noted (p. 51). The beaver is only found west of the River Yenisei, in the streams of north and mid-Siberia ; but it has grown very rare, and at the present time has almost disappeared. It has a commercial value not only for its skin, but for the castoreum (beaver's cod) obtained from it. The *ush-kan*, or Siberian hare, is spread over the country : it is grey in the summer and white during the winter, and has very long hair. About 1,000,000 or 1,200,000 skins are sold a year, half of them at Irbit. The Siberian peasants neglect it as a means of food, and only unwillingly eat its flesh. The polar pika hare comes as far south as lat. 47° N. Other rodents are the vole, hamster (west of the Ob), jerboa (south of the Steppes), and hairy-nosed porcupine (at greater altitudes). The rat is a great pest, but its activities are somewhat restricted by building store-houses on poles.

Sea-mammals

The sea-mammals of Arctic Russia and western Siberia are not nearly so important as in eastern Siberia. The seals found are the true seals (*Phocidae*), in both west and east, but the sea-bears or fur-seals (*Otaridae*) only occur in the east. True seals have a great commercial value by reason of their hides, their blubber, and the oil produced from their fat. But they are extremely destructive of the fish : in many places where the seal appears, the fisherman abandons his occupation, knowing that his task is useless on the arrival of the *kozhya* ('leather :) as he calls him. In places like the Kola Peninsula they ought to be kept down much more energetically, for the protection of the fishing industry. At Ponoi the seal-hunting in March and April is very profitable. In the Kara Sea there are no seals; so that region is without attraction for polar bears. Seals ascend the Amur as far as Eri in about lat. 51° N. In Lake Baikal and up the Selenga is found a distinct species, *P. baicalensis*, which is hunted for its skin.

The fur seal (*Otaria ursina*) occurs in great numbers in parts of the northern Pacific, including the Commander Islands, Robben Island, and the Pribilov Islands ; there at the end of May or early in June the males arrive : then about the middle of June arrive the females, and give birth to their young. The pups are black when born, but in August have a fresh coat of grey fur. The promiscuous killing of fur seals when in calf out at sea led to the protracted disputes in connexion with the Bering Sea arbitration. The sea lion (*O. Stelleri*) visits the Pacific coasts in autumn. No fur seals occur in the Arctic Ocean.

The walrus is found in the islands of the Arctic Ocean, and at various points west of the Lena (between which and the Bering Strait it is never found), e. g. Ponoi and the Kara Sea. It no longer occurs in the Sea of Okhotsk. It is differentiated from fur-seals by the absence of external ears, by the structure of the teeth, including the presence of tusks, and by its more substantial build. It is hunted for the sake of its blubber, its hide, from which are manufactured harness and sole leather, and its tusks, the ivory of which, however, is far less valuable than elephant ivory. Its breeding season is from April to June.

The chase of the bowhead or right whale has decayed in the Sea of Okhotsk and Bering Sea.

The finner whale (*Balaepoptera*) is found along the coasts of Arctic Russia and eastern Siberia. The whale 'fisheries' of Finmark are active, but those of the Kola Peninsula are feeble. There were two for a time at Port Vladimir and Ara Bay, but they closed for lack of capital. During their short period of activity they killed 300 whales. A well-developed whale fishery might do great business, but it might have to face the antipathy of fishermen in other parts of the coast on the same grounds as those on which the Norwegian fishermen attacked the whalers. These reasons were, first, the whales were thought to drive away the Greenland seal which preys on fish, and secondly because whales feed on capelan, a kind of salmon which pursues the cod inshore, so that the destruction of whales reduces the number of cod. The beluga or white whale (*Delphinapterus leucas*) is found in these waters. Its skin is cut into broad strips and sold to the Samoyedes and Yuraks for reindeer harness.

Birds

There are said to be 285 species of birds in the whole of Siberia, few of which are unknown in Europe. The chief line of demarcation of species is thought to be the watershed between the Yenisei and Lena, but so much of the country east of that boundary is unexplored that it is difficult to be very precise on the subject. Seebohm, in his book *The Birds of Siberia*, has given a great deal of information about the birds of Arctic Russia and the valley of the Yenisei.

The Arctic zone has few permanent residents: only the ptarmigan, snow-bunting, raven, snow-owl, and Icelandic falcon are found there always, but with the return of summer the tundra becomes full of bird-life, and it is the breeding-place of a vast variety of birds from the beginning of May. They are bountifully provided by nature with berries that have been frozen throughout the winter, and with swarms of mosquitoes. The commonest of these summer visitants seemed to Seebohm to be the Asiatic golden plover. But few birds stay longer than the beginning of winter, when they migrate to the belt of coniferous taiga further south.

Edible birds.—It is difficult to say what a native of Siberia will or will not eat, and it is probable that his taste in the flesh of birds is as catholic as in other forms of food. It is enough, therefore, to mention the most important of the birds that can be eaten. Duck of various kinds, including a long-tailed species, are common over all Siberia, especially in the lake district north of the Kirghiz and Baraba steppes; around Narim they assemble at different lakes: they are frightened from one to another, their way being netted in advance. They are thus caught and then killed by having their necks bitten through. They are found in great numbers along the lower course of all the Siberian rivers. Geese are found in multitudes as far north as the mouth of the Yenisei. The goose falls an easy prey in the moulting season, when it cannot fly. Among other edible birds are ptarmigan, willow-grouse or kuropatka, swans, hazelgrouse, pronounced as especially delicate eating, capercailzie, blackcock, partridge, and heathcock. In the neighbourhood of Biisk and Novo-Nikolaevsk, woodcock, hazelhen (or *ryabchik*), and quails are common from May to the end of August, and during the threshing season 5,000 birds, mostly quails, are netted. In hunting the capercailzie a barking dog is used, which has an irritating effect on the bird; the hunter then gets in a shot, while its attention is distracted. The same practice is adopted in hunting the elk.

Birds valuable for plumage.—Most valuable of all is the eider duck, found along the shores of Arctic Russia, by reason of the down that is obtained from it. But the dwellers along the Pomorski coast are killing off these ducks for the sake of their flesh and pillaging their nests for eggs. Grebes are of economic value: crested grebes used to be shot in great numbers in the Tyukalinsk district in order to make muffs. Magpies, which are particularly frequent in the steppe district west of the Ural, are of value for the feathers, which are exported. A thousand are taken annually in the Biisk district.

The pigeon for religious reasons is never molested among the Russians. The seagull enjoys a like immunity for the same reason among the Samoyedes.

FISHERIES

A. Arctic Russian Fisheries

Speaking generally, there are three main kinds of fishing on the north coast of Arctic Russia: (1) In the open sea there is fishing for cod, haddock, flatfish, and wolf-fish. (2) Along the coast there is fishing for herring and *navaga.* (3) Up the rivers there is fishing for salmon and certain freshwater fish, including also salmon-fishing along the coast. There is some hunting of sea-mammals also along this coast.

Murman Coast

Cod.—The main fishing here is for cod (*Gadus morrhua*), the yearly average catch being 10,000,000. The fishing-season is from the end of March till the end of August. As the cod comes from Norway, the fishing begins on the west coast first and works eastward. It employs the local inhabitants, and about 3,000 men from the districts of Kem and Onega. The cod-fishing is done in small sailing boats (*shnyaks*), which are worked by four men and a boy, who has to roll up and dry the nets: this crew is called an *artel*, the name applied to what corresponds to a trade union in Russia. The men who come from the White Sea are exploited by *procruters* or factorists, who supply them with their vessel, their tackle and supplies for themselves and their families, in return for which they are entitled to one-third of the catch, but the value of the goods supplied is deducted when accounts are squared, and as the valuer both of these goods and of the fish caught is the *procruter* himself, the fisherman finds himself more and more in his power.

The course of the warm Atlantic waters varies, so that the

voyages are regulated by the existing currents. Usually the fish are at something between seven and twenty miles from the shore and at a depth of from 120 to 150 fathoms. One way of fishing is with lines, 180 fathoms long, but this needs a great deal of muscular exertion, and is not adopted except by poor fishermen. The other and more effective method is to use 'gartlins' or great lines (*yarus*), which are sometimes five miles long, made of rope of the thickness of the little finger with finer ropes of from 3 ft. 6 in. to 4 ft. 8 in. in length attached to it at intervals of about 2 ft. 4 in. About 5,000 hooks are attached to a *yarus*, and these are baited with capelan (a small oily fish like a smelt), or sand-eels, sandworms, or the inside of crabs. The *yarus* is lowered to the bottom of the sea and taken up again after six hours. Over 1,200,000 hooks are used each year on the Murman coast.

Other fish caught on this coast are turbot (*Rhombus maximus*), haddock (*Gadus aeglefinus*), coal-fish or saith (*Gadus virens*), wolf-fish (*Anarrhichas lupus*), flat-fish (*Pleuronectidae*), comber or sea-perch (*Serranus cabrilla*), eelpout (*Lota vulgaris*). The herring-fishing on this coast is practically disregarded, though herrings are numerous. The Greenland shark is the inveterate foe of the whale, and whales have been captured which show traces of combat with this relentless enemy. Sharks come in pursuit of cod, and can be caught with a line on the Murman coast, but despite the value of shark-liver and the oil extracted to adulterate cod-liver oil, the fishing is only casual, and not properly developed. From those sharks whose skin is not too rough is made shagreen for smoothing or polishing wood. The shark is never eaten, on the ground that it is a great eater of human flesh itself; and it is an enemy to the fisherman, because it despoils his *yarus*. At certain places on the west coast of the Murman are fat-melting works for the fat of sharks and stock-fish.

Prices of fish.—The price per pud of fish on the Murman in the last decade of the nineteenth century was as follows: cod 10*d.* to 2*s.* 1*d.*, haddock 2½*d.* to 10*d.*, wolf-fish 3¾*d.* to 8¾*d.*, coal-fish 2½*d.* to 7½*d.*, turbot 2*s.* 3*d.* to 2*s.* 6*d.* The prices for fish products were cod-liver 1*s.* 3*d.* to 2*s.* 6*d.*, shark-liver, haddock-liver, and coal-fish-liver 10*d.* to 1*s.* 5½*d.* Another fish product is the fish-guano made of the discarded portions of fish, especially torn and dried fish-heads. The total value of the Murman fishery in 1897 was £48,158.

Winter employment of fishermen.—An auxiliary trade for the fisherman to carry on during winter is carpentering, e. g. they manufacture the wooden packing-cases in which the fish are exported, and find it very remunerative.

There are government salt depôts along this coast.

White Sea

The main fishing is for salmon, herring, and *navaga*, which is done in home waters, and is more especially the work of the old men, women, and children, while the more enterprising and active men do deep-sea fishing for cod or go seal-hunting, starting out for these employments early in the spring.

Salmon.—The salmon-fishing lasts from the middle of May till the beginning of November, but the best salmon are caught in August and September when the salmon come up the rivers to spawn. The fishing is done at the estuaries of the bigger rivers by means of dragnets, trammel-nets, &c. Higher up the rivers it is done by *zakots* or weir traps : close to Kem is a *zabor* (a sort of fence trap) made of logs embedded in the bottom, in which a large net (*morda*) is inserted. Salmon of large size (about 22 lb.) are thus caught. The spring-catch is for local consumption, the fish caught early in the year having less taste than the others ; the autumn catch is sold to dealers.

Herring are caught along the Pomorski coast mostly from the end of June till October, but the fishing continues in winter. To the north-west of Kem the best fisheries are Kandalaksha, Keret, Kovda, and Knyaz Bay, from which about 25,000,000 fish are obtained in the year. The fish are salted and shipped to Arkhangel in barrels of 27 lbs., but these are so badly put together that the fish deteriorates and only realizes a low price. To the south-east of Kem the principal fisheries are Soroka, Shizhnaya (from each of which the annual catch is about 5,000,000), Sukhona volok (with an annual catch of 2,000,000), Virma with an annual catch of 1,500,000), and Sumski (with an annual catch of 1,000,000). The herring is caught with poke-nets, sweep or drag-nets, &c. The whole catch is sold on the spot to dealers, the fish being either frozen or smoked. The usual price is 1*s*. to 3*s*. per 1,000. A fish loses its value if it is frozen and then salted; so the winter-catch is eaten fresh. In the neighbourhood of Sumski there are several smoke-drying sheds, in which 5,000,000 herrings are annually smoked.

Navaga.—The *navaga* (*Gadus navaga*) is a fish allied to the cod. The season for fishing is from November to January, and it is done through holes in the ice. It is so easy that it is an occupation usually left to the children. The *navaga* is caught chiefly by the inhabitants of Kolezhma, Sumski, Shuya, and Nyukhcha. Those caught at the first two are about twice the size of those caught at Shuya, and average about 1 lb. 14 oz. A load of good *navaga*, consisting of about 4,000, costs from 16*s*. to £2. In February the *navaga*, having spawned, becomes lank and tasteless: in March it migrates.

Other White Sea fish are cod, flat-fish, wolf-fish, gwyniad (*Coregonus lavaretus*), but these are not an object of export-trade, but are only for local consumption. There are two fish also, the lumpsucker (*Cyclopterus lumpus*) and father lasher (*Cottus scorpius*), which are not eaten by the inhabitants, but dried and given as food to the cattle.

The Karelians, especially at Pongamskaya and Keret, hunt marine animals.

Gulf of Mezen to Kara Sea

The fishing along this coast is of much less importance. The industry is almost entirely in the hands of the Samoyedes. Owing to the absence of markets they deliver their catch to monopolists. The Russian dealers supply tackle, &c. to the natives, whom they organize in small groups on a system that closely resembles that of the *procruters* with the Pomors.

Marine animals are hunted, when they are brought down on floating ice by the northerly winds: among these are the walrus, Greenland seal, sea-hare (*Phoca leporina*), and white whale (*Delphinapterus leucas*). The seal blubber is sent to Arkhangel, thence to Russian and foreign markets. Along the Gulf of Mezen during August and September there is fishing for salmon and *nyelma*, and in November and December for *navaga*. The waters round Kanin Peninsula and Kolguev Island were for a long time untouched, despite their vast supply of cod and flat-fish, but lately very large plaice have been caught by British trawlers and other foreign boats. About a hundred steam trawlers in these waters make an aggregate haul of 40,000 tons, chiefly plaice.

Lake and River Fish

In 1897 there were caught 1,192 tons of lake and river fish, valued at £13,112. The chief districts in which they were caught were Pechora, Kem, and Kola, the best salmon being found in the Northern Dvina, Onega, Mezen, Pechora, and Varzuga. In Lake Imandra in the Kola Peninsula Engelhard caught salmon, grayling (*Coregonus thymallus*), gwyniad, trout, and salmon-trout. The *nyelma* (*Coregonus leucichthys*), a salmon with white flesh, is found in these lakes and rivers.

In the Pechora are found the peliad (*Coregonus peled*), the *omul* (*C. omul*), a salmon with white flesh, of primary importance farther east, and the *chir* (*C. nasutus*). The *Coregonus* is the characteristic salmon of the Arctic Ocean, as the *Oncorhynchus* is of the Pacific. In the Dvina is the sterlet (*Acipenser ruthenus*), a smaller member of the sturgeon tribe and the one which penetrates highest up the rivers.

B. Western Siberian Fisheries

The Ob Basin

The basin of the Ob constitutes the whole of western Siberia, and all the fishing centres are either along its waters or along those of its confluents or the lakes from which its waters are fed. The chief are on the lower waters of the Ob and Irtish, on Lake Zaisan and the upper Irtish, and on Lake Chani near Kainsk, the fishery of which is stimulated by the immediate contiguity of the Siberian Railway. There are other places where there is fishing, as in Lake Marka-Kul and the lakes of the Kirghiz steppes, but in these it is a secondary and subordinate occupation of the inhabitants. In the Ob 42 species of fish are known, the most abundant families being the Cyprinidae with 15 species, and the Salmonidae with 12; among others are three species of sturgeon (the sterlet, the *sturio*, and the *ossetr*), two of perch, two of cod, and pike is extraordinarily abundant. There are but slight differences between the fish of the Ob and the Irtish.

Regions of the Ob basin.—Varpakhov divides the Ob basin into three regions—lower, middle, and upper—differentiated to some extent by their species of fish.

The first region includes the Ob estuary and extends as far as Berezov. The characteristic conditions of the region are masses of water with a very slow current or stationary, and abundance of 'sands', i. e. stretches of clear water with sandy or rocky bottoms. The chief fish of this part are chub, *navaga*, *seld* (*C. merki*), a member of the salmon tribe with white flesh, salmon, *pidchian* (or *sig*), *chir*, *muksun*, a broad fish with large bright scales and small head, sturgeon, pike, roach, *gremille*, peliad, and eelpout, some of these being common to the whole river.

The second region extends from Berezov to about lat. 54° N., up both the Ob and the Irtish and their tributaries, coinciding approximately with the wooded territory. The type of fish found especially in this region demonstrates the abundance of lake and marsh, e.g. crucian carp, tench, sterlet, and other fish. Salmon and *muksun* are found in great numbers here as well.

In the third or upper region which extends from lat. 54° N. to the sources of the Ob and its confluents the characteristic fish include trout, grayling, *seld*, *Gobio fluviatilis*, and *taimen* (*Salmo fluviatilis*) which gives its name to many villages. The *nyelma*, a very popular fish in Siberia, is the commonest in the whole river. Of course there are no very marked boundaries between these three zones: sometimes, for instance, sterlet, which is characteristic of the central region, descends to the Ob estuary, or grayling, a typical fish of the mountain streams, is found in the tributaries of the lower Ob. Some fish may be regarded as local, which always inhabit the river, while others, which are migrants, visit it from the Gulf of Ob or cross from one locality to another.

Migration of fish.—The movement of great masses of fish takes place everywhere after the ice breaks, especially about the middle of May. These movements do not take place simultaneously with the various kinds of fish. The first arrival is the peliad, followed in order by the *muksun*, the *nyelma*, the *pidchian*, the *chir*, the sturgeon, the eelpout, and last of all the *seld*. They ascend slowly, making about 40 miles a day. Many of them (the *muksun*, *nyelma*, and peliad) make for the *sori* (shallow backwaters which dry up in hot weather), and then when in early autumn there is later abundance of water they move up the river and spawn in its higher waters and in those of its upper tributaries. Others, such as the *pidchian* and *chir*, spawn in the lower Ob, and the *seld* only comes into the lower Ob and its tributaries. The sterlet spawns in flooded meadows. The sturgeon goes up to the upper Ob and spawns on the way: in the autumn some sturgeons remain in the river, but more go down to the sea where they are caught in great numbers by the natives in the Gulf of Taz. Sometimes the grampus or killer-whale arrives in the lower Ob: its appearance is the signal for the fish to leave the river and go up the backwaters, where an occasional grampus will follow them.

The Zamor.—There is one special phenomenon of the Ob that deserves attention, as it has a great influence upon the fishing. It is called the *dur* or *zamor*, or 'the dying of the water'. In December and January the accumulation of protoxide of iron brought down by the tributaries of the lower

Ob that flow through tundra and marsh, coming over the lower surface of the ice, proves fatal to the existence of all living things. It gives the water an unpleasant taste and smell. The process goes on unequally, first in the shallow parts, and reaches from the middle of the stream to the shores. In rapid and deep places the *zamor* does not exist, and the fish do not die. The instinct of the fish makes them escape before the *zamor* can destroy them. Pike, roach, and *nalim* go into tributaries where the waters are not so corrupted, sturgeon and sterlet to the mouth of the Ob, and some of the sterlet up the Irtish. The *zamor* gradually extends up and down the river, and by the end of the winter the central and lower regions of the Ob are lifeless. On the Irtish this phenomenon is found up to the village of Semeika. The effect upon the fishing is obvious; the great mass-movements of the fish towards the sea are obstructed by fish-dams. Instinct forbids retreat up stream, and multitudes of fish are accordingly caught.

Methods of capture.—The favourite instrument of capture is called *gimga*. It is like the *morda* of European Russia, but larger. It is made of long thin twigs, on a substructure, which costs at least £100 to erect, so that it is a method only within the reach of those who are possessed of capital. The fish trying to escape the massive obstructions come to the gaps where the *gimga* is set, and fall ready victims. The *gimga* is so close-woven that even quite small fish are caught by it. The number of *gimgas* at various points of the river differs in accordance with the breadth from 40 to 100. The part where the majority of them is set is between Berezov and Obdorsk; 500,000 fish are sometimes caught in one day. In the lower course of the Irtish and in the Ob near its estuary instead of *gimgas*, there are set with the obstructions *cherdaks*: these are four-cornered sacks made of netting, attached to long poles by means of which they are lowered and raised. These methods are employed in the summer fishing as well as in winter, and on a smaller scale by local fishermen when the fish are leaving the *sori*. A good deal of fishing is done with nets:

big nets 4,200 ft. long and 70 ft. high, or half-nets more simply constructed and of smaller dimensions. These are employed generally on the ' sands '. In the neighbourhood of Tobolsk there are 120 of these ' sands '. The big traders employ the first kind, the half-nets being used by those of humbler means. A large amount of ice-fishing is done especially in the reaches of the Irtish between Tobolsk and Semeika (where the *zamor* stops). The ice is divided into sections, and snares armed with hooks are let into the water through the holes that are made in it. In some places 800,000 hooks are let down, but the catch is not great, varying from ten to four hundred puds.

The fishing industry.—About 10,000 men take part in the fishing industry of the Ob. The poorest are the impoverished Samoyedes and Ostyaks, men who have lost their reindeer and taken to fishing : an epidemic among the reindeer always adds to the number of fishermen. But, with the exception of the Reindeer Samoyedes and a few fur-hunting Ostyaks, all the inhabitants of the uncultivated north along the Ob are engaged in fishing. The great centre of the industry is Tobolsk on the Irtish, which is the head-quarters of the six big fishing firms that erect the largest dams. Every spring, as soon as the ice clears, the summer expedition proceeds down the river from Tobolsk. They give pay on an average of about 30 roubles a month and provide certain supplies : at the beginning of October they return to Tobolsk. About 1,000,000 puds (15,000 tons) are taken annually, of the value of between £300,000 and £400,000. 50,000 puds are taken annually to Irbit fair in February from the middle Ob. Farther north the winter catch is kept till the summer and sold to the summer-fishing expeditions. Fish are transported by sledge, a weight of about 20 puds to each sledge, the transport industry necessitating the existence of a race of winter-dwellers along the Ob. Sterlets are caught in winter near Tobolsk, fetching 5–8 roubles the pud in that city. Good sturgeon are caught on the river, weighing 8 puds and containing half a pud of caviar. The sturgeons of the Ob are much bigger than those of the Irtish. Pike are sometimes so numerous on the Ob that they are sold for only 12 kopeks, but this is not surprising as a company of 50 or 60 can take in one season up to 7,000 puds. 350,000 puds of fish are carried annually on the Siberian Railway, 200,000 on the Perm-Tyumen Railway.

Centres of fishing industry.—Besides on the Irtish below Tobolsk, and the Ob below the mouth of the Tom, fishing is one of the chief occupations of the population along the Rivers Om, Tom, and Chulim, and higher up it is of importance to a large number of inhabitants of the Altai district, not only on the Ob between Barnaul and Biisk, but on tributaries like the Kondom in the Kuznetsk district. No statistics are furnished, but there is a generally prevalent local belief that the fish are on the decrease. On the Ob itself it has been calculated that the fish exported from the different regions are as follows: Obdorsk 300,000 puds, Berezov 150,000, Samarovskoe 75,000, Surgut 90,000, Narim 90,000.

Fish products.—Several canning factories have been erected. Caviar is obtained from the roe, and isinglass from the bladder of the sturgeon. Besides these two important products, the sturgeon is of importance for its fat, dried sturgeon being fatter than smoked salmon, and for its spinal cord, which is eaten raw or else dried and cut into small pieces and used for baked fishcakes with fish inside and dough outside, while it furnishes a constituent in *selanka*, a soup which is the Russian national dish. *Poziom* is prepared from sterlet, *sirok*, and *muksun*: the fish is split open, freed from bones, salted, dried in the air and slightly smoked. In summer the fish taken from the Ob are dried and salted, in winter they are frozen.

Lake fisheries in Western Siberia

Lake Zaisan is the centre of a considerable fishing industry, partly in the hands of Cossacks, partly in the hands of Kirghiz. Carp, trout, *nyelma*, and sterlet are found, but roach and perch are more numerous. The fishing begins at the end of April and continues till the end of August. The height of the fishing season is June, after which the fish begin to go down the Irtish. Pavlodar on the Irtish is in a district where fishing is very important.

Lake Chani.—This lake has the great advantage of being served by the Siberian Railway. About 100,000 puds are exported annually. The principal fish are pike and crucian.

The *River Ural* is practically outside the sphere of this book, but is important for its protected fishery, especially for the sturgeon.

C. Eastern Siberian Fisheries

The principal fisheries in eastern Siberia are the lower Yenisei; Lake Baikal and the rivers that flow into it; the Lena and other rivers of the Yakutsk Government; the Okhotsk and Kamchatka fisheries of the Pacific; the Amur and its estuary; and the coast of the Ussuri Province (south-west fishery).

The Yenisei Basin

As an important industry the fishing of the Yenisei is practically concentrated in its lower waters. The fish most sought in these waters are the sturgeon, the sterlet, the *nyelma*, the *omul*, the *muksun*, the *seld*, the gwyniad, the *chir*, and the *sig*. Most of these make long migrations up and down the river to spawn; many sturgeons stay in the deep pools of the river, at any rate during the winter; they begin to go up the Yenisei when the ice melts, at the end of May or the beginning of June. A good many fish stay in the estuary throughout the winter; some fish, like the sterlet and *chir*, keep to the river all the year, and are never found at its mouth. Altogether, about fifteen varieties are caught for the purposes of trade, including the sturgeon, sterlet, *muksun*, *nyelma*, and *omul*. So that the caviar may retain its quality, the sturgeon is often kept alive in floating fish-tanks; it is said that the *omul* which migrate are fatter than those which stay in the estuary all the year.

The fishing is done partly by Yenisei-Samoyedes, Yuraks, and some Dolgans, Tungus, and Ostyaks, partly by non-resident Russians. The work of the natives is exploited by

Russian buyers, who, as a rule, do not give money, but goods on credit, a system which leads the fisherman into perpetual debt. The Russians who live along the river fall victims to it as well as the natives. The natives, who own fishing-places, usually let them cheap and act as fishermen. Primitive methods are employed in preparing the fish, with little regard for cleanliness, so that a rotten smell is a constant accompaniment of fish from the Turukhansk district ; the preparation of caviar is equally primitive.

Every year, at the beginning of June, boats containing the necessities for the season are rowed or towed down the river, reaching Dudinka in about three or four weeks, dropping fishermen and supplies at the river-stations as they go. The traders buy some of the natives' winter catch, and reach Yeniseisk again at the beginning of August. After ten days they start on their second voyage, and about the beginning of October the expedition is back again at Yeniseisk with the men and the summer catch. The first voyage only brings back strongly salted fish, the second brings back what is less strongly salted ; some of it is dried. What they cannot carry is often brought by sledge to Krasnoyarsk during the winter. For salting, as on the Ob, steppe salt is used ; the proportion is usually about 180 lb. of salt to 700 lb. of fish.

In the rest of the Yeniseisk Government the fish trade only amounts to about £5,000 a year. About 3,000 puds reach Minusinsk from the upper Yenisei. Lake Bozhe, in the Achinsk region, also produces a certain amount of fish. But most of the fishing is only for local consumption.

The fishing in the Yenisei is done principally by seines, with, as a rule, five men to a net. The big employers of labour usually make combinations of two or three, and have a tug ; they completely control the smaller workers. The rich men have nets of nearly 20,000 ft. in length, but the natives have to be content with much smaller nets. Besides nets, there are also dams of interlaced branches stretched across the river. In the winter some fish, especially sturgeon, *omul*, and *muksun*, are caught under the ice. Hunger and curiosity make them fall at this period ready victims to any bait.

Absence of good communications has greatly restricted the fish-industry of the Yenisei. There is no canning, and the attempt to send frozen fish by rail to Russia has been a failure, despite the demand for such supplies. The present amount exported south annually is about 175,000 puds, including about 155,000 puds of summer-salted fish and 20,000 puds of winter-salted fish, but the market is almost entirely confined to the Yeniseisk Government, with Yeniseisk and, to a lesser extent, Krasnoyarsk as centres of the trade, though a few of the fish from this region go as far as Tomsk and Irkutsk. The best fish are found a long way north, and it is only possible to make one voyage within the year for the summer catch.

About 175,000 puds are caught in the Yenisei and the shallow tundra lakes by the local population and used for the needs of themselves and their dogs. The annual value of the Yenisei fisheries, including export fish, is about £80,000.

Lake Baikal

The Baikal fishing-region includes Lake Baikal itself, the lower reaches of the rivers that feed it, especially the Barguzin, the Selenga, and the Upper Angara, and the lagoon-like lakes along the shores of Lake Baikal, termed *sori*. The principal fish of these waters are the sturgeon (which is found in the Lower Angara and Lake Baikal, and fished for in the latter during the winter through the ice), the *omul*, the *chir*, the gwiniad, the grayling, the roach, the crucian, and the burbot. There is also a mysterious fish, the *golomyanka* (*Comephorus baicalensis*), which lives only in the profoundest depths of this lake, and is about 10½ ins. long. In Lake Baikal there is further a species of seal (*Phoca baicalensis*). In other parts of the Transbaikal Government are found, besides most of the fish of Lake Baikal, pike, carp, tench, and silurus. In Lake Frolikha, near the north extremity of Lake Baikal and communicating with it by a river of the same name, is a special kind of trout, not known elsewhere (*Salmo erythreas*).

The main fishing in the Baikal fisheries is for the *omul*, of which 500,000 are taken yearly of the value of about £20,000. During the winter the *omul* keeps in the deepest waters of the lake; in the spring it begins to approach the shores and enter the small inlets along them; towards September it moves in masses to the estuaries, up which it goes to spawn; it ascends the Upper Angara for more than 60 miles: at this period the 'running-catch' is made. Later, when it has spawned, it goes back to the lake, and the 'swimming-catch' takes place. When the lake is frozen, it is caught under the ice by nets let down to a depth of 100 to 150 fathoms. The winter catch is put on the market frozen, the summer and autumn catch salted.

The implements for taking fish in Lake Baikal greatly vary, nets and 'bagnets' being used where the fishing is on a large scale. Bagnetting is carried out by small companies; there are also net associations, where each member supplies a settled number of fishing-nets and ropes.

Lena and Kolima

The Lena and Kolima region is of very little industrial importance owing to the absence of means of communication, but none the less a great deal of fishing goes on to satisfy the needs of the inhabitants. For most of them fish is the staple food, and has the same importance that grain has for the inhabitants of agricultural districts. The natives eat chiefly small fish. Thus 94 per cent. of the inhabitants of the Yakutsk district are engaged in fishing, and it is the occupation of 92 per cent. in the Kolima district, and 87 per cent. in the Verkhoyansk district, and of 68 per cent. in the Olekminsk district. It is possible that the Kolima fishers will find an outside market for their fish, as since 1911 there has been regular steamer communication with Vladivostok. The amount of the catch of the district, including the adjoining lakes, is estimated at from 4,000 to 5,600 puds. At present the only market for the Lena fishery is the mining district.

The chief fish are sturgeon, sterlet, *muksun*, *nyelma*, gwyniad, *chir*, bass, common gremille, dace, pike, and burbot. A great number of herrings are found in the estuaries of both the Lena and the Kolima. Crucian carp is specially common. Fishing is most vigorous on the lower reaches of both rivers. About 25,000 puds are exported annually by steamer from Bulun up the Lena to Yakutsk. The Aldan, with its tributary, the Maya, is also prolific in fish. The fishing on the Kolima is vigorous, so far as the conditions allow, but the river is frozen for 268 days in the year. The fishing on that river is mostly done by companies, but, even though clubbing together, they have very insufficient implements. At ninety-nine fishing-stations along its lower waters there were only fifteen entire nets in all, the remaining fishermen contenting themselves with broken parts. The methods of preparation are as inadequate as the fishing-tackle ; caviar is hardly prepared, and is often thrown away, as the natives do not eat it. Frozen fish is frequently eaten like cheese, cut into thin slices and called *stroganin*. The sturgeons are very large, often weighing as much as 200 lb. The coast dwellers hunt for seals, especially in March and April.

Okhotsk-Kamchatka

This district in the Pacific extends from Udskaya Bay, where the River Uda flows into the sea (in lat. 55° N.) to the mouth of the River Anadir (in lat. 65° N.), taking in the coasts of the peninsula of Kamchatka and of the Commander Islands. It is divided into a western and an eastern section by Cape Lopatka, the southern point of Kamchatka.

In the Pacific the conditions of the industry and the species of fish are entirely changed. Instead of a Russian monopoly there is keen competition with Japan. In fact the Japanese had got the fishing trade almost entirely into their hands, before the Fishing Convention was made in 1907, which gave them free rights of fishing, except in certain specified bays and river mouths. Even now a great deal of fishing is under Japanese control, and the market for the fish is largely

Japanese. In the western section only one bay (Penzhina Bay) is excluded from the convention ; in the 1,850 miles of coast in the eastern section sixteen bays and gulfs are excluded, but yet in only one of these (Avacha Bay) is the industry carried on by Russian enterprise. Several areas have been closed to all fishing since 1913 in order to conserve the fisheries. These include the mouths of the Ulya, Urak, Okhota, Kukhtui, Kola, Tau, Yana, Arman, Ola, Yama, Takhyama, Nayakhan, Gizhiga, Tigil, Bolshaya, Osernaya, Kamchatka and other rivers.

The fish of the Pacific differ largely from those of the Arctic Ocean and of the rivers that flow into it. The characteristic Salmonidae are not *Coregoni*, but *Oncorhynchi*. The principal salmons of the Pacific that ascend the rivers that flow into it are six in number. (1) The *chavucha* (*S. orientalis*), confined to Kamchatka and the Sea of Okhotsk, the largest of the tribe, but a fish that has not yet established itself in European markets ; it is a fine fish with good flavour, averages 15 to 20 lb., and is often six feet long. It supplies the best caviar, experiments showing that the best comes from fish over four years old. (2) The *goltsi* (*S. collaris*), a kind of sea trout, ascends the rivers to the head waters, and returns in the following spring. (3) The *keta* (*Oncorhynchus lagocephalus*) or dog-salmon, is the commonest of all in these waters, except in south Kamchatka (where the *chavucha* is most prevalent) ; it weighs nine or more pounds ; a man can catch 1,000 in a day. It is of inestimable importance to the natives ; its skin provides them with sails, dress, and boots ; it is preserved in various manners, and forms the chief food of the inhabitants of Primorsk ; its caviar, which is of a pale red colour, is now regarded as of value, though previously it was thrown away. The *keta* is a very timorous fish and avoids clear water : it comes in great shoals. (4) The *gorbusha* (*O. proteus*), the humpbacked or Alaskan pink salmon, is less choice : it weighs from four to eight pounds, or occasionally even ten. It is found in all the rivers. (5) The *krasnaya* (*O. lycoodon*) or red salmon is smaller than the *chavucha* and appears a fortnight after it. It weighs eight pounds, and salts better than other species. (6) The *kizhucha* (*O. Sanguinolentus*) is most plentiful about the beginning of August. It weighs four or five pounds.

Herring and cod, though numerous, are not articles of trade. Only in Gizhiga and Penzhina Bays, when there is scarcity of salmon, do the inhabitants make use of herrings for themselves and their dogs. Yet they advance on the east coast of Kamchatka, when the ice breaks up, so closely packed that they can be dug out, it is said, with a spade, and towards the beginning of June they almost block certain places like the mouth of the Kukhtui River near Okhotsk. There are a great number of cod also about the Commander Islands, but the natives do not touch cod, and the trade in them is entirely in the hands of Americans, who salt them for Japanese and Chinese consumption. The cod are so numerous that sometimes, when breeding, shoals of cod reach a length of 1⅓ miles and a depth of several feet.

There are other fish also, the *mikisha* (*Salmo purpuratus*) largely a Kamchatkan fish ; the *valek*, a fish of Okhotsk ; the *korzhukha*, a kind of smelt, and the *uiki* (*S. socialis*), which frequent the west coast of Kamchatka, where there are hardly any herrings, but of all these the natives take no stock. Farther north, in Anadir Bay, there are pike, grayling, dorse, and malma, as well as salmon, but, though there has grown up a fishery that aims at commercial value as well as the satisfaction of local needs, it is difficult for it to be exploited properly at that distance. About 350,000 are caught ; the fish goes mainly to Japan, which receives two-thirds of the total, amounting to nearly 10,000 tons ; some also goes to Vladivostok and some to Petrograd direct. The inhabitants of Markovo secure about half a million red salmon, in addition to the white. The local population makes large stores of dried fish (*onkoli*) for their own food and that of the dogs.

So far as local consumption is concerned, there are two periods of fishing in the year : (1) in spring, when the hungry inhabitants, who have had difficulty in eking out their winter supply, feed themselves; the earliest arrivals being plaice, haddock, and shad, and the western Kamchatkan coast being visited before the north coast of Okhotsk; and (2) the summer fishing, which has to supply the winter needs of their dogs and themselves; this begins in mid-June in Kamchatka.

Commercial fishing is largely in Japanese hands. For the whole Far East the fishing-stations are put up to auction every year at Vladivostok by the Department of Domains. The total amount realized in 1913 was £31,419; the amount three years before was only £9,700. The number of Russians that take up the fishing-stations is increasing; in 1910 only 5 per cent. were taken by them, in 1912 22 per cent. In 1913 there were 148 stations in western Kamchatka, of which only 9 were Russian, and 61 in the eastern Kamchatka region, most of which were Japanese. There were also 39 Russian river stations in Kamchatka. There is a strong tendency in Kamchatka to look for better customers than the Japanese, who beat down prices. But the cost of freight makes it almost impossible for the fish of this region to compete with those of the Amur. Again all labour and supplies have to be brought from Vladivostok. Not only is the distance from Japan considerably less, but the workmen are paid less, their food costs less, and they have a large supply of schooners and steamers for handling the fish. The only communication between these uninhabited districts and Vladivostok or Nikolaevsk is by sea during a short sailing season by the limited steamer service of the Volunteer Fleet, and all nets and material have to be brought a long distance. The Japanese fishermen secure abundant supplies from Hakodate by their own vessels.

Canning has been started in Kamchatka with some success; the work is mostly done by Japanese firms. The chief canneries are on the Osernaya, Bolshaya, Kamchatka, Palana, and Kolpokara rivers. In 1913 the total output of tinned salmon in Kamchatka was 137,314 cases of four dozen 1lb. tins.

The extent of the salmon fishery in Okhotsk and Kamchatka can be realized from the numbers of fish taken in 1913, which were: Okhotsk, 500,000; western Kamchatka, 33,500,000; eastern Kamchatka, 11,800,000. The herring catch in these districts totalled about 188,000.

Besides this salmon in 1913 there were prepared in Okhotsk 213 tons of caviar, in western Kamchatka 1,134 tons, in eastern Kamchatka 1,034 tons.

There are various ways of preparing fish: one, called *yukola*, of a crude nature, is only applied to fish intended for the consumption of natives or dogs; the form of fish-preserve which is most exported is called *balyk*. It is exported from Okhotsk to Vladivostok and Yakutsk, and from Petropavlovsk and Ust-Kamchatsk to San Francisco and Vladivostok.

Seal Fisheries of Commander Islands

One special marine industry is the hunting of the sea bear or fur seal, which supplies 'sealskin'. The centre of this, as of the cod industry, is the Commander Islands, where the creatures congregate in the summer. In the course of the last twenty-five years they have greatly diminished owing to immoderate fishing in the open sea. In 1890 no less than 55,435 reached the market, but in 1911 only 200. In 1912 a prohibition against killing them for the next five years came into force. During this period there is every reason to hope that the breed will have been regenerated and restored, especially as hunting these animals in the open sea has been prohibited by the Washington International Commission for fifteen years. With the renascence of the sea bear it is hoped that the Commander Islands will recover in prosperity and population.

Amur

Fishing districts.—There are three fishing districts in this region: (1) Nikolaevsk, the most important, comprises the lower Amur for 200 miles above its mouth, the Amgun, the Amur estuary, about 130 miles of the coast of Sakhalin, and about 860 miles of the coast of the Okhotsk Sea from Udskaya

Bay to the Amur estuary. (2) Mariinsk, from Troitskoe to Sofiisk, a reach of about 263 miles. (3) Khabarovsk, above the last district as far as Khabarovsk, a reach of about 107 miles.

In the Nikolaevsk district there are three kinds of fishing villages: fish-catching stations, salting stations which buy but do not catch fish, and villages which do some fishing incidentally. Leaving out of account the many villages in the last category, in 1913 there were 111 fishing stations leased from the Government at a total annual rental of £32,000 and 28 fishing stations leased from the municipality of Nikolaevsk at a total annual rental of £17,000. In the Mariinsk district there were 27 and in the Khabarovsk district 3 commercial fishing stations. In the two latter districts all the stations were Russian. Other stations were given free of charge to certain villages in order to ensure their food supply. The fisheries of the River Ussuri are entirely in the hands of Cossacks or natives who fish for their own needs with primitive methods. Prongé, lying south of the Amur mouth, used to have an important fish trade with Germany. The Volga caviar merchants have a station in this region which was reported to be very successful.

Japanese fishing.—Up to 1899 the Japanese invasion so completely monopolized the fishing industry that hardly anything remained over for the Russian population or Empire; in that year foreigners were prohibited from fishing in the Amur and its estuary, and Russians were forbidden to use foreign labour. Under these conditions development of the fishing industry became possible, and it has been especially stimulated since the Russo-Japanese War. In 1907 a convention was concluded with Japan by which the Japanese were admitted to the same fishing rights as the Russians. From this convention all rivers and thirty-four bays in the Far East fisheries were exempted, and it is practically in these alone that Russian fishing prospers. In the Amur estuary foreigners are allowed to prepare but not to catch the fish. The chief buyers in the Amur fishery were originally the Japanese, but they have lost their market by trying to force down prices. The catch in 1910 was so good that it enabled the trade to send great quantities to Europe, and the business thus inaugurated has continued. In 1912 the fish trade with Japan from the Maritime Province was practically extinct.

Amur fishery.—The conditions of the fishing industry on the Amur have greatly improved. The fact that better prices now obtain ought to stop the excessive fishing that formerly prevailed, but probably stringent regulations will have to be introduced. In the remoter districts the rule that fish may not be caught within two versts of a river mouth is constantly disregarded. A fish hatchery is to be established at Nikolaevsk by the Government, which is alive to the danger of the rapid exhaustion of the fisheries under the present method. With a view to encouraging local consumption the experiment is to be made of giving fish a prominent place in the rations of the troops. The Amur stations are being equipped according to the latest plans with refrigerators, electric-light installation, and all processes that make for cleanliness.

Salmon.—The most important fishery is for salmon, and then for sturgeon. Besides the fish specially named there are about fifty varieties of less commercial value. The salmon of the Amur are the *keta*, which enters the river from the end of June, and the *gorbusha*, which ascends the river at intervals from the middle of August to the middle of September, often going up-stream 1,200 miles. On the lower Amur the average weight of the spring *keta* is 4 lb., of the autumn *keta* 9 lb., and of the *gorbusha* 2½ lb. In 1913 the catch on the lower Amur, exclusive of the Nikolaevsk district, was about 1,340,000 *keta*. In the Nikolaevsk district the catch was about 18,260,000 *keta*, and 7,500,000 *gorbusha*.

In 1913 there were sent 46,031 tons of fish and fish products by rail via Khabarovsk. The trade has become so much stimulated that special storage-houses are to be built at Moscow and Vladivostok. The fish for the Russian market is either frozen or salted, and sent in 25-pud barrels ; for Japan it is dry-salted in the Japanese way, without barrels. There is a rapidly growing trade in salmon-caviar, especially the roe of the *keta*, which used formerly to be thrown to the dogs, but 3,652 tons of it were in 1912 carried by railway in refrigerators. There is some trade also in train-oil made of fishes' livers, about 10,000 gallons being secured in a month and a half. There is only one canning factory in this region : in 1913 it turned out about 100,000 tins of salmon, each of 1 lb.

Scientific investigation has revealed a good many facts about the Amur salmon. It is a migratory fish, that lives in the sea and ascends the river only at spawning-time. The young fish make their way down to the sea and live there for three or four years, after which they assemble in large shoals, and ascend the river against the current for more than 1,200 miles. In their life as river fish their colour and appearance change. After spawning they become weakened and are swept down by the current, while so many die of exhaustion that there is a general belief that all fish that enter the river succumb.

Sturgeon.—Sturgeon-fishing on the Amur is mainly a winter industry, but there is serious danger of the fish being exterminated. The Government have had little success in attempting to limit the season from June 15 to the melting of the ice and to prevent the use of drag-nets. Fishing goes on all through the year, even during the spawning season. Large specimens are already scarce, the average being from 30 to 40 lb. In winter they are caught by hooks through holes in the ice in large pools. which the natives know them to frequent. The consumption is almost entirely local. In 1913 the total catch was 147 tons. The Government imposes a tax of $\frac{3}{4}d.$ per lb. The better kind is caught in the neighbourhood of Khabarovsk, though the sturgeon is much more plentiful near Nikolaevsk. A certain amount of caviar is shipped to Vladivostok. One special form of sturgeon on the Amur is the *kalug*, or white sturgeon.

Sakhalin

The fishing in Sakhalin is losing importance. Here besides the *keta* and *gorbusha*, which swarm in August, herring is a common fish ; it is used almost entirely as fish-manure, which is exported to Japan. In 1913, a bad year for herring, the total catch in the fourteen stations in Russian Sakhalin was about 200,000 salmon and 4½ million herring. About 274 tons of herring manure and an equal amount of salted salmon were exported to Japan. Smaller quantities of fish and caviar went to Russia. The Gilyaks engage in herring-fishing, when the *keta* season is over. To the Gilyak fish is the principal form of food. His supply for winter is almost exhausted by December, though there is fishing for dorse in Baikal Bay during the winter. Then comes a time of great hardship. In April the seals appear, but before their arrival comes the haddock, which is hooked through holes in the ice. Then come herring and halibut (*Pleuronectes hippoglossus*), which sometimes weigh more than 100 lb. Trout (*Salmo fario*) appears in the rivers, but the next great catch is the ide (*Idus melanotus*), which is caught by baskets in the rivers. The smelts (*Osmerus eperlanus*) are so numerous that they are often ladled out of the water. An ally to the Gilyak fisherman is the grampus or killer-whale, a voracious animal which drives fishes and seals before it up the rivers or on to the coast ; in return for these services the natives give it a friendly reception if they meet it alive, and inter its body with due rites if it is washed ashore.

The Gilyak have a special type of weir or dam for catching fish ; this, as well as a Japanese weir, is constantly used also on the mainland.

The South-west

The sea-coast of the Ussuri district of Primorsk is known as the south-west region. It extends from the boundary of Korea to Cape Lazarev at the south of the Amur estuary. In this region seven bays, including Peter the Great and Imperatorskaya Bays, are excluded from the convention.

North of Peter the Great Bay the industry is principally in Japanese hands. The reservation of the fishing-rights in this bay for Russians has put an end to Chinese and Korean trade, and the proximity of a good market in Vladivostok has greatly helped Russian industry. Steam-trawling is beginning in this bay, the first trawler being British-built. Fish caught in this way are salted.

The chief fish caught in these waters is herring; it approaches the shores and enters Peter the Great Bay about the middle of November, comes again through the end of December, January, and the beginning of February, and pays a final visit in April. The principal herring fisheries, however, are north of Imperatorskaya Bay. Besides the herring, the *keta* and *gorbusha* are also obtained. Other fish, like the smelt, flounder, mackerel, and dorse, are of much less economic importance. Counting the three most important fish, the catch in 1913 was about 41,000,000 herring, 658,000 *gorbusha*, and 138,000 *keta*.

An important fishery is that for trepang (*Holothuria edulis*), the Chinese name for the Golden Horn of Vladivostok, applied to the *bêche-de-mer* or sea-slug. It is found on rocky bottoms along the whole coast of the Primorsk, but is especially common in the neighbourhood of Peter the Great Bay. The Chinese spear or net it. There are two seasons, from the end of March to June and from mid-September to October. The average catch for a fisherman is 120 a day. They weigh about six pounds when dried. In 1913 about 9 tons were exported from Vladivostok.

In 1913 about 125 tons of dried crabs went to China and Korea. There is a considerable market for shrimps.

Products of Wild Animals

Fur

Fur is the oldest established trade of Siberia. Originally the lure of the invaders, it was for many years the form in which the subject tribes paid their tribute. Siberia is extremely rich in fur-bearing animals, but with the growth of civilization and the destruction of the forests, they are being driven further and further north, the number of many breeds is being seriously diminished, and rigid restrictions on their capture have to be introduced, if they are to survive. As a rule, the colder the climate, the better is the quality and colour of the coat. The lower the latitude, the less silky is the fur, and the hair is apt to be 'harsh' in the tropics, lacking in softness and depth.

In western Siberia the most important parts for fur-bearing animals are the Berezov, Surgut, Turinsk and Obdorsk districts in Tobolsk, and the Narim district of Tomsk. In certain parts of Tobolsk Government hunting is the chief occupation of a large section of the population, providing them with the means of existence, e. g. along the Rivers Vakh and Yugan; for the inhabitants of the valleys of the Rivers Agan and Torum-Yugan it is as important as fishing, and for those of the lower Ob it often supplements fishing and reindeer-breeding. In the Narim district of Tomsk it is one of the main industries, as much as fishing, for the Ostyaks, and is an addition to the industries carried on by Russian settlers. In the Kirghiz steppes there is hunting, but it is principally for sport; it is only the poor who trade with the proceeds of the chase, though they do secure a certain number of wolves, foxes, and ermines.

The chief fur animals found in western Siberia are, in the Tobolsk Government, the squirrel, fox, ermine, and hare, and in the Berezov tundra the Arctic fox; less common are the kolonok, sable (diminishing greatly in the Turinsk district), brown bear, wolf, and, beyond Obdorsk, the polar bear, on the shores of the Arctic Ocean; in Tomsk, of most importance are the squirrel, kolonok, sable (greatly decreasing), fox, ermine, bear, hare, and skunk.

In Yeniseisk hunting is the chief occupation in the north, and also in the Turukhansk and Angara districts. Elsewhere it is only supplementary. The deer, the Arctic fox, the hare and the squirrel are the chief animals. The others are rare and small. In the Irkutsk Province the natives are nearly all trappers. The Kirensk and Verkholensk districts were once well stocked with animals, but they are rapidly declining. Nearly all the fur animals are found here, the squirrel being the commonest. In Yakutsk hunting is still the principal occupation of nearly all the inhabitants, but the decrease in the game is rapidly making it secondary to fishing. However, fur is still the basis of barter with the natives. In the Amur and Maritime Provinces hunting is the chief means of livelihood for the natives, but in the Amur Province not for the Cossacks. In the Amur the natives are chiefly occupied with hunting the sable for its fur, while the Russians mainly hunt the roebuck for its leather. A number of valuable animals are found in the Maritime Province. (For a detailed description of the fur animals, see pp. 52–59.)

The great centre of the trade in Siberia is the Irbit fair, though the December fair at Ishim also has importance, especially for the sale of squirrels' fur. But besides these big centres there are many smaller fairs arranged at the close of the hunting season, which is usually the early winter. The hunting is often done by co-operative groups, who share profits. The fur trader often deals with these people, directly trafficking in things like tea, sugar, tobacco, gunpowder, and manufactured goods, which are indispensable to the inhabitants, for the furs which they have come to buy. The smaller fairs are losing their importance. But the fur trader is often the agent of a bigger man, and the fur passes from hand to hand till it reaches one of the great fairs. The agents of the larger firms, principally German and English, push farther and farther inland.

It is difficult to give accurate figures of the number of animals killed, and the amount of fur secured; we do not know how much is used in local consumption, in manufacturing clothes for the inhabitants, but the figures of Irbit fair are accurately known, and furnish the best evidence of the present state of the trade. Irbit fair takes place from February 8 to March 10. It is supposed that two-thirds of what is for sale there comes from the country west of Lake Baikal. What is not sold at Irbit usually goes on to Nizhne-Novgorod. A great deal of sable is not sent to Irbit, but direct to Moscow. Now, owing to the restriction on killing sable (mentioned on p. 54), which may be continued, sable appears to a much smaller extent among the sales at Irbit. The most important centre and distributing point of the fur trade is London, the next most important is Leipzig, near which place, at Weissenfels, there is a gigantic industry in dressing the skins of Russian grey squirrels and making them into linings. Irbit used to be eighty miles from the railway: the opening of the branch that goes through Irbit and Turinsk from the Perm-Tyumen Railway to the Tavda may do a great deal to recreate the greatness of the Irbit fair by making it more accessible; but it looks as though the importance of fairs would be lessened, as the traders are more and more making their purchases at the place of production and dealing with the trappers on the spot.

In 1910 the total sales at Irbit amounted to £729,000; in 1912 prices had risen from 15 to 20 per cent., and the total amounted to £833,000; in 1917 with a great increase in prices the sales totalled £724,160. The sales in 1912 included 4,535,000 squirrel-skins, 1,500,000 rabbits, 12,250 sables, 200,000 ermines, 1,500 brown bear, 180,000 kolonok, 16,500 grey wolf, 14,000 to 15,000 fox. Very fine sable sold for £42 each; black fox skins were scarce, and fetched anything from £21 to £105; grey wolf skins were sold from £1 5*s.* to £3 11*s.* In 1914 the total value of squirrel-skins alone was £228,000, but in 1915 there was a great drop in the sale of these to £90,000. In 1916 business was rather slack. The chief fur-sales were as follows: about 3,500 badgers; 1,500 bears, the price being about £3 4*s.* 5½*d.* per skin; 60,000 black cats fetching up to 1*s.* 5½*d.* per skin; 10,000 Orenburg marmots; 1,000 pine martens from £1 12*s.* 2½*d.* to £1 18*s.* 7½*d.* each; 500 stone martens, £1 7*s.* 11*d.*; 6,000 mink; large quantities of red fox; small quantities of Yeniseisk white fox at £3 4*s.* 6*d.*; 7,000 Obdorsk white fox, ranging from £2 5*s.* to £3 per skin;

some Pechora white fox at £2 13*s*. 8*d*.; about 100 silver fox, ranging from £16 2*s*. to as much as £107 each; about 4,000 reindeer fawns sold up to 7*s*. 6*d*., the same number of summer reindeer skins sold from 10*s*. 9*d*. to 15*s*. 0½*d*.; only some 3,000 sable, the lowest price being £2 15*s*. 10*d*., the highest £8 11*s*. 2*d*.; squirrels were very numerous, and fetched high prices, the best with full tails (3,000,000), realizing from 8*s*. 7*d*. to 9*s*. The Chinese bought up most of the dark ones on the spot, and only about 500,000 of them were offered for sale, less valuable types fetching 10*d*. or 1*s*.; 1,000,000 white hares sold up to 1*s*. 1½*d*. for the best; about 250,000 white pole-cats from 3*s*. 5*d*. to 3*s*. 9*d*.; wolves were much in demand, about 2,000 were sold, fetching £1 18*s*. 7*d*. The best from Turukhansk sold from £3 0*s*. 1*d*. to £3 15*s*. 1*d*.

Fur-fairs are held at the confluences of the chief tributaries of the Amur, at Albazin, for instance, and Blagovyeshchensk. In the Uda district, where the sable is especially good, the fairs are at Kulcha on Lake Orel, Burukanskaya on the Tugur, and at the mouth of the Uda. There is an important fur-fair, on a much larger scale than these, at Nikolaevsk. As the natives of this region remove the heads and claws of the bears from religious motives, the bear-skins here are not of much value.

In the far North-East the Anyui fair, once of the first importance, has declined considerably, since the Chukchee prefer to barter most of their furs with the Americans on Bering Strait. Only the most valuable furs are sent to the Anyui, as for these Russians give better prices. A number of furs, walrus, and mammoth-tusks now reach Vladivostok by sea from Gizhiga. The Russian traders at the Anyui fair all come from Yakutsk.

At one time the Kyakhta fair, where Chinese merchants bought the peltries, was of great importance, but this is now no longer the case.

There are many small fairs throughout the north and east. The fur sold at those in the south goes to Yeniseisk and Irkutsk. But the great fur-market of the north is the Yakutsk fair, which is held in July. The following table shows the number of skins sold :

	1902.	1905.	1913.	*Price.* £ s. d.
Sable . . .	2,640	3,000	—	
White fox . .	8,400	14,000	20,000	(in 1911–12) up to 2 14 0
Fox . . .	844	5,000		
Red fox . .			1,000	
Grey fox . .			100	2 12 1
Kolonok . .	220	1,000		
Ermine . . .	900	12,000	10,000	5 9
Squirrel .	73,500	300,000	70,000	1 2
Black bear . .			100	

In the average of the total amount of skins of fur-bearing animals in the international markets, the share of Russia and Siberia is as follows :

	All Russia.	*Siberia.*
Squirrels	15,500,000	15,000,000
Hares	5,250,000	5,000,000
Ermines	1,100,000	700,000
Kolonok	200,000	150,000
Skunk	300,000	150,000
Brown bear . . .	8,000	6,000
Sable	215,000	70,000

A word or two may be added about the last-named animal. Its numbers had fallen so alarmingly that a law has been passed forbidding its slaughter from February 1, 1913, to October 15, 1916, with a permanent close time from February to October in each year. The Moscow Fur Association has pressed for a renewal of this law for another three years. In Kamchatka the danger of the sable becoming extinct was recognized some years ago, and reserves were marked out within which the hunting of sable was prohibited. A recent expedition has staked out two large sable-reserves in the Sayansk district. Sable skins range from 15 to 20 in. in length, and from 5 to 8 in. in breadth. In genuine sable the outer covering of hair is especially delicate in quality and beautiful in colour, having a rich blue tint, and varying from 1½ to 2½ in. in length, while the pelt is very soft, but durable. In Kamchatka, Sakhalin, the Maritime Province, and the Barguzin district of the Transbaikal the sable holds the first place. It is, indeed, the most valuable of the fur animals. The best black sables come from the Yakutsk Province, notably the Lena district, the lightest and least valuable from the Ob and the Yenisei. The Kamchatka sable is browner than the others. The fur of the kolonok (' Tartar sable '), which is really yellow, can be dyed so as to resemble sable with such success, that expert judges often cannot tell the difference.

Another animal for which protection is necessary is the white fox. Though it is found throughout northern Siberia, it is in great demand for imitating the rarer black and silver fox furs, and is ruthlessly hunted in consequence.

The Indigirka is now the centre of the white fox hunting, the skins being bought up by the agents of the Ust-Yansk and Yakutsk merchants. In 1911 good skins fetched from 15*s*. to 30*s*. there.

The squirrel appears in the largest numbers in the fur-markets. The farther north and east the animal is found, the darker, thicker, and more valuable is its fur. Next to the squirrel comes the hare. The best squirrels come from the Lena, but the ermines from that region are the least valuable.

Tiger-hunting is a regular occupation in winter on the lower Amur. In 1912 fifteen were killed and twelve caught alive. Of these ten were sent to Hamburg for sale. In some years 120 or more are killed. The bile, heart, and claws are sold to the Chinese, who make from them a powder, which is supposed to produce courage.

Game

The export of game is small. In 1909 it was about 1,200 tons, worth £53,000 ; in 1910 about 1,800, worth £73,600. But Siberia abounds in edible birds of many kinds—duck, geese, hazel-hens, ptarmigan, &c.—and the export might be much increased if the business were better organized, and more ice-wagons were supplied on the railways. The principal place for the export of wildfowl in western Siberia is Barnaul.

Fossil Ivory

The collecting of fossil ivory is a regular industry among the natives of the far north. The mammoth tusks are found principally along the Arctic Ocean. Those near the shore are usually smaller than those inland. The New Siberia Islands are a favourite hunting-ground, the waves washing the tusks out of the sand-dunes in stormy weather. On the mainland these are most commonly found embedded in the earth banks of the smaller streams, the spring floods exposing them to view. The natives make their way to the Arctic Ocean or the adjacent islands with their dog-sledges in April, returning in November, when the ice is firm again. The Yakutsk fair is the chief market for fossil ivory, which is little inferior to ordinary ivory. Nearly twenty tons were taken in 1913, the price at Yakutsk being about £5 10*s.* for 36 lb. A certain amount is brought to the fairs in the Tobolsk Government. The Chukchee possess beautiful breast-ornaments made of fossil ivory. The flesh of the mammoth is eaten by the natives, apparently without harm to themselves, but Europeans will be loth to follow their example.

CHAPTER V

NATIVE TRIBES OF SIBERIA AND ARCTIC RUSSIA

Classification: I. Palaeo-Siberian Tribes. II. Neo-Siberian Tribes: (i) Finno-Ugrian, (ii) Samoyedic, (iii) Turkic, (iv) Mongolic, (v) Tungusic

Classification

THE inhabitants of Siberia may be considered in three groups, corresponding to the different periods at which the country was settled. (1) The first group are the descendants of the pre-historic inhabitants, who have always existed in the country, or else entered it at an early period of which for Siberia there is practically no knowledge. Such are the Chukchee, the Koryak, the Gilyak, and these are the aboriginal inhabitants. (2) Secondly there are races who settled in Siberia during the great movements of population which took place in Asia from the third to the thirteenth century. Such are the Finno-Ugrian tribes—for instance, the Vogul, and perhaps the Samoyedes—who came from the south of the Altai Mountains in the third century ; such the Turkic tribes, Yakuts and Tartars, who came from the regions of the Oxus, perhaps in the eleventh century or earlier ; such the Mongols, a people akin to the Turks, coming from the regions of the Himalayas into Siberia in the time of the famous Jenghiz Khan, early in the thirteenth century ; such are the Buryats. (3) Thirdly there are the Russian colonists, who have come more or less continuously into Siberia ever since the notable expedition of the Cossack Yermak, in 1580.

The tendency to expand eastward is very distinct in Russian history. With regard to Siberia, Russian historians are fond of comparing the stream of immigrants to a military column, which has thrust itself along a broad belt of territory stretching east and west, from the south of the Ural Mountains to the valley of the River Amur. The territorial belt thus settled, more densely than any other part of the country, begins on the west, between Verkhoture to the north and Troitsk to the south, and stretches eastwards between Tobolsk and Petropavlovsk; between the Tara and the Om the belt narrows, broadens a little between Tomsk and the Biya, and, as a compact mass, ends about Nizhne-Udinsk, with the rivers which flow into the Yenisei. In its eastward march from the Urals the column of immigration has thrown off outposts, as it were, which form settlements along the Ob, Yenisei, and Lena; and gradually the Amur valley is being penetrated. By this more or less thickly settled belt of colonization, the native races have been divided into a northern and southern portion; but far the greater number of them lie to the north, along the valleys of the great rivers that flow into the Arctic Ocean.

A few native stocks are said to have disappeared altogether, or to have been absorbed; some, like the Gilyaks and Chukchee, in the far corners of the land, are practically intact; some, like the Irtish-Ostyaks, have preserved their nationality even amidst Russians; some by intermarriage have exercised a greater influence upon the Russians than the Russians have upon them; this is especially true of the numerous tribes of the Yakuts, whose customs, dress, and even language, have to a great extent been adopted by the immigrants. This mixing of Russians with native stocks is commonest on the outskirts of the belt of settlement, along the Yenisei and Lena. It is the converse of what happened, in the Middle Ages, in Russia itself, when native stocks from central Asia invaded the country, and intermarried with and were absorbed by its inhabitants. Only where the immigrants are comparatively thick in Siberia, do the natives become Russianized.

In this way new types have arisen in Siberia; in addition to pure native stocks, there are native stocks infused with Russian elements, stocks whose dominant element is Russian, but who have been distinctly affected by native elements; and finally there are the pure Russians, who, from living in the novel conditions of life, social and geographical, of Siberia, have themselves become a new type, different from the western Russian.

It is difficult to make a satisfactory classification of the tribes of Siberia. The name, Palaeasiasts or, better, Palaeo-Siberians, has been applied to those indigenous stocks, which bear no clear relation to the other races that inhabit the world, but their own mutual relations are very indeterminate. To the rest, sometimes called loosely the Ural-Altaic stock, is given the name of Neo-Siberians. The word Mongolian is used too loosely for the purposes of scientific classification, sometimes so as to cover all the yellow races, including the Palaeo-Siberian tribes; at other times for the Ural-Altaic stock, so as to include together Finns, Tartars and Turks; again at other times, and more accurately, it is applied to one branch of this stock, i. e. that of which the Kalmuks and Buryats are members.

Palaeo-Siberians.—To the Palaeo-Siberians can be assigned without further definition of their relations to one another the following not very numerous tribes: Chukchee, Koryak, Kamchadal, Gilyak, Yukaghir with their branch the Chuvanzi, Ostyak of Yenisei; outside Siberia are the Ainu, Aleut, of whom some live in the Commander Islands, and Eskimo, of whom a small number have returned to Asia.

Neo-Siberians.—The Neo-Siberians can be divided into five fairly clear branches: (1) the Finno-Ugrian, which is practically confined to western Siberia and Europe; (2) the Samoyedic, which extends along the north coast beyond the Yenisei, and has one branch, the Soyots, among the Sayansk Mountains; (3) the Turkic; (4) the Mongolic, whose great representative is the Buryat; and (5) the Tungusic, which includes the Manchu and many of the tribes of northern Manchuria.

Altogether the number of these natives in Siberia is roughly a million. Many of the tribes are dying out; few only, like the Yakuts and Buryats, are increasing in numbers and have set their impress upon the Russians who have settled among them.

In the more barren parts for instance among the marshes of the Irtish, or in the regions towards the Arctic Ocean, where there is little or no scope for agriculture, and where hunting and fishing are almost the only occupations, the number of natives is not increasing. But in the more fertile parts of the country, in the best parts of the river-valleys, in the middle and south, the native stocks do tend to increase. It is in these parts also that the process of Russification is most clearly taking place. Generally speaking, it may be said that the annual number of births among the natives is satisfactory; diminution in their numbers comes from a disproportionately high death-rate, particularly among infants. Nominally most of the natives are Christian, and belong to the Orthodox Greek Church. But the old superstitions and the old gods to a great extent prevail among them, while the more outlying tribes have hardly been touched by Christianity at all. They still have their images, magic, and medicine-men, their sacrifices, and their taboos; polygamy and loose morality are not uncommon; in places, even slavery exists; the blood-feud is handed on from generation to generation. Amid cold and dirt, they live lives which to the western mind appear to be of unrelieved discomfort, bringing with it painful diseases, unrelieved by any medical help, and only aggravated by the brandy for which they greedily offer their wares.

I. The Palaeo-Siberian Tribes

Chukchee

Territory.—The Chukchee territory proper lies east of Chaun Bay and north of the Anadir. But now, owing to their growing herds, the Chukchee have spread even as far as the Indigirka in the west, over the Anyui tothe Omolon, and to the River Opuka and the Polpol Mountains on the Pacific. The few who reached Kamchatka have been largely assimilated by the Koryaks (see p. 106). Their territory consists chiefly of tundra with a fringe of forest, the camps lying mostly along the rivers, which are separated by bare watersheds. In autumn they seek the edge of the forest for shelter, in summer the hills near a small glacier, or preferably the open tundra. A camp wanders about 100–150 miles, following the same track each year. If the ground proves unsatisfactory, another may be chosen, so long as it is not already occupied. It is worth noting that by entering the Anyui and Omolon territory the Chukchee have put an end to the migrations of the wild reindeer between the Omolon and Chaun Bay. Hence the Yukaghir (see p. 113), who live almost entirely upon them, are rapidly dying out.

The villages of the Maritime Chukchee stretch from Cape Erri to Anadir Bay, except for a few Eskimo settlements.

Name.—The word 'Chukchee' is generally derived from *cau'cu* (rich in reindeer). It is the name by which the Reindeer Chukchee distinguish themselves from the Reindeer Koryak or the Maritime Chukchee. But both Reindeer and Maritime Chukchee call themselves *li'i-yi-lulit* (those of genuine language) as distinct from other tribes.

Language.—The Chukchee language is very similar to that of the Koryak. Though rich in words and pliant, it is less vital than Koryak and it has virtually no dialects. The similarity of the Chukchee language, as of their stature and features, to those of the Indians on the other side of Bering Strait is said to be noteworthy.

The dominant position of the Chukchee is shown by the way in which they force other tribes and even the Russians to speak their language. Even on the Kolima and the Anadir they speak very little Russian.

Numbers.—The Chukchee probably number about 12,000. The greater part are reindeer-breeders, who inhabit some 650 camps. Thanks to their success with their herds they are increasing steadily and are the most powerful tribe in the east of Siberia.

Relations with the Russians.—Since the failure of their attempts to subdue the Chukchee in the eighteenth century, the Russians have left them virtually independent and the relations between them have been, on the whole, excellent.

Many parts of the peninsula have never been visited by a Russian.

The power of the Chukchee chief whom the Russians recognize is barely nominal. Since 1889 a tribute of 247 roubles is paid by the Chukchee at the Anyui fair, but most of it comes from wealthy reindeer-breeders, and apparently the Russians are again obliged to give very substantial presents to secure it, now that the fair is so rapidly declining.

Tribal divisions.—There are two great divisions of the tribe, Reindeer Chukchee and Maritime Chukchee. There is considerable evidence, apart from tradition, to show that the Chukchee were at one time entirely a maritime people which only turned its attention to reindeer-breeding by degrees. Their folk-tales are generally of the sea and the dog figures more largely in their religious life than the reindeer, though it has long ceased to be used by the Reindeer Chukchee for driving and is kept chiefly for religious reasons. Some of their stories point to their having migrated from the south, as do their names of the months. The process of transformation is still going on. Many of the Maritime Chukchee have a few deer in the herds of friends and thus gradually acquire the means of starting as breeders. Even the Eskimo are following their example, for the depredations of American whalers are rendering the livelihood of the Arctic people who live on sea-mammals more and more precarious. Not that reindeer-breeding is without its risks. In a bad winter many owners may lose half their stock. But it is more profitable and more stable, on the whole, and many coast villages have already ceased to exist.

Social organization.—The family is the permanent Chukchee unit, but the camp is the economic unit. It generally consists of a few families, perhaps of ten to fifteen persons in all. Rich people generally divide their herds, forming new camps. Permission to join a camp must be granted by its members. Each camp has its master, who is also called 'the strongest man', and lives in the 'front tent'. The maritime villages only occasionally have such a master. The commoner unit with them is the 'boatful' of eight rowers and a helmsman who commands, and whose nearest relatives form the crew. The catch is divided among them on a regular system. The 'strong man' has more influence in the village than in the scattered camp. Murder within the family group is dealt with by the family alone and dangerous or disagreeable members are sometimes done away with. Murder outside the family entails a blood-feud; for revenge is a duty. A group of kindred families is called *varat* (i. e. 'collection of those who are joined together'). But this bond of union is nowadays very loose.

Physical appearance and characteristics.—In appearance the Chukchee are well-built and healthy though heavy, and the well-fed reindeer-breeders are wonderfully strong. The nose is large and well-formed, but the lower part of the face is disproportionately heavy. Except on the Pacific coast, the hair is black. That on the face is scanty, but a moustache is the sign of manhood. The skin of the Maritime Chukchee is darker than that of the Reindeer. Women are more often of the Mongolian type, but many are as fair and shapely as the average of the white race. On the whole the stock is pure.

Marriages with Russianized creoles are perhaps increasing, owing to the prosperity of the Chukchee, but they are generally childless. There is a marked decrease of syphilis in the present generation owing, perhaps, to the precautionary measures taken in the last. But the country is liable to be swept by epidemics of measles and other diseases. The Chukchee are easily angered, but quarrels are usually settled by fights or wrestling-matches. Murders are still fairly numerous, however. The language is notably poor in terms of abuse. The kindness of the Chukchee towards suffering, even in animals, is most noticeable, as is their gentleness with children. They are also wonderfully generous towards other tribes. In time of famine a rich breeder near the Kolima will kill hundreds of animals for hardly any return. Some of the Lamuts (see p. 103) on the Chaun tundra get half their food from the Chukchee. Though there is now no ill-feeling towards foreigners, the Maritime Chukchee are more hospitable than the others and less given to stealing. The Reindeer Chukchee are continually robbing each others' herds, the theft being punished by a fine or a thrashing, where the victim is strong enough to enforce his rights. The Chukchee are slow-witted and easily cheated in business. Their quinary-vigesimal system of counting is clumsy in the extreme and they can only keep track of the more notable animals in their herds. They have no remedies against disease except magic. Their endurance of cold is astonishing. Women will sew in the open snow half naked, because the exertion makes them so hot. They are not clean. They even call themselves the 'non-washing people'.

Fishing.—Seal-hunting is the chief occupation of the Maritime Chukchee. They use light harpoons for stabbing the seals through their blow-holes in winter, when they do not net them. For stalking them in the open they use heavier ones, but these are being rapidly superseded by guns. Walruses are much less common, though since the Americans no longer hunt them they are a little more numerous. On the Pacific coast they are most easily killed during their migrations between Kresta Bay and Cape Dezhneva. Walrus and white whale are the favourite food. The Chukchee skin-boats are made out of one or at most two walrus-hides. They are light and can carry more than a whale-boat, but they are easily holed. Whale-boats are being more and more used, but owing to the scarcity of wood they are difficult to make.

Hunting.—The Reindeer Chukchee, in addition to breeding reindeer, also hunt the wild reindeer when they cross the Anadir between the mouth of the Main and Chikayeva. They leave the Polpol Mountains in March and continue crossing in June, beginning to return in July. Some twenty Chukchee families on the middle Anadir live on nothing else and each gets 100–200 deer in a year. Non-migratory deer are also shot and mountain sheep are highly prized. Wolves, bears, and white or red foxes are trapped, while birds are snared. North of the Anadir there is comparatively little fishing.

Bows and spears are still used in quarrels and every Chukchee wears a knife on his hip. The Chukchee make themselves snow-goggles out of leather or wood, with narrow slits for the eyes. The armour of walrus-ivory, seal-hide, or iron is now kept only as a curiosity. The Chukchee dogs are poor, though the Maritime Chukchee eat them in time of famine. The excellent Anadir dogs fetch high prices among the Chukchee.

Dwellings.—The Chukchee tent is large and round with an oblong inner room which is the chief habitation. The three central poles have a sacred character. The tent is always set to the same points of the compass and the left side belongs by custom to the master. The inner tent is lit by a single lamp and the main evening meal is eaten there. Guests strip to the waist, while the family is naked except for a belt, as the heat rapidly becomes stifling. The stench is intolerable. Older and thinner skins are kept for the summer tent.

Clothing.—The chief garment of a Chukchee is a heavy double loose-fitting reindeer-fur shirt, the collar of which can be tightened with a string. His boots and trousers are also double and of the same material. In these and his cap he can sleep in the open in winter. The Maritime Chukchee buy the cast-off clothes of their Reindeer brethren, who never wear them two winters running. In very bad weather they also wear a cape or a long great-coat. The women wear long boots and clumsy combinations, the sleeves of which so interfere with their work that they frequently keep one arm and breast bare. The Chukchee woman tattoos very little.

Food.—The Maritime Chukchee live largely on sea-mammals, the Reindeer Chukchee on meat, but each at times craves for the other's diet. The Reindeer Chukchee are not squeamish. They will eat meat or entrails in any state of decomposition. They drink large quantities of tea and all the alcohol they can get and are inveterate smokers. Like the Koryak, they make an intoxicant from a mushroom.

Birth and marriage.—The Chukchee are prolific, many families having from 5 to 9 children.

Marriage is not permanent. A man may change his wife. As a rule, however, the marriage is broken by her relatives reclaiming the bride. There are also 'group-marriages' of 10 couples, in which the husbands have a right to each other's wives. But this tie is never made between people in the same camp. On the death of a husband his brother succeeds him, keeping the dead man's reindeer herd for his children. Polygamy is rare among the Maritime Chukchee, as they cannot afford to support two wives, but not uncommon among the Reindeer breeders.

Death.—Chukchee funeral rites are largely a protection against the evil influence of the dead. The body is drawn up through a hole in the roof or the back of the tent and all traces are removed to prevent the dead man's return. It is taken to the burial-place on a sledge. Here it is opened, the organs examined, and the cause of death proclaimed. The throat is then cut. The corpse is either exposed or burned. It is afterwards visited, to see whether it has been carried off by beasts—the best sign. On the fifth day the tent is moved to another place, but sacrifices are afterwards made at the grave. The usual abode of the dead is thought to be underground. Those who die sudden or violent deaths dwell in the Aurora Borealis.

Religion.—*Vairgit* among the Chukchee are the benevolent beings to whom sacrifices are made, and they live in the 22 directions of the Chukchee compass. The chief one lives in the zenith; and Midday, the Sun, and the Pole-Star are very important. Others live in the reindeer and the walrus and in the winds. There are three classes of *kelet* or evil spirits, (*a*) invisible spirits, bringing disease and death; (*b*) bloodthirsty spirits, the enemies of warriors; (*c*) spirits which assist the *shaman*. The *kelet* is fond of the liver. Hence the opening of a corpse to discover what *kelet* has attacked its liver. According to the Chukchee there are from five to nine worlds one above the other, connected by a passage under the pole-star. Other parts of the sky are also inhabited.

The object of Chukchee ceremonial is to maintain the

welfare of the community, and incantations are the leading feature. The chief festivals are the autumn and winter slaughterings, the ceremonial of antlers and the sacrifices to the New Moon, the Fire, and for Luck in Hunting. Moreover, each family must perform a thanksgiving ceremony twice a year.

Sport.—The Chukchee are fond of sport. They will travel enormous distances to race their reindeer in the spring. There are also foot races and wrestling matches.

Reindeer-breeding.—The Chukchee reindeer herds, which are probably the most numerous in the world, are the most important economic feature in far north-eastern Siberia. It is therefore convenient to include in this section a general account of reindeer-breeding in eastern Siberia with more particular reference to the Chukchee.

The Chukchee laid the foundations of their present prosperity in their raids on the Koryak herds during the eighteenth century, but the principal increase has been during the last fifty years. The son of a chief who used to be looked on as very wealthy because he owned two herds, possesses five to-day, while his brother-in-law and his cousin each own three.

The Chukchee deer are imperfectly tamed and readily run wild again. Milking is out of the question and they are difficult to manage in harness. The breed is undersized with short head and legs, heavy body, and thick antlers, and is dark in colour when compared with the Lamut. It is good for food, fattening quickly and keeping its condition. But it cannot be ridden and is much weaker than the Lamut, which is of twice its value. A Lamut fawn is exchanged for a full-grown Chukchee deer. Hence the Chukchee use Lamut deer in harness and sell their own to the Lamut for food. Crossing between wild and tame animals is common, the wild deer visiting the Chukchee camps in the rutting season. The fawns are much valued for racing, as they are swifter and have more mettle than the others. Their pedigree is preserved for three or four generations. A cross between a wild doe and a tame buck is especially prized. In colour the deer vary from dark grey to hazel, the fawns being darker than the full-grown animal. They live from twelve to fifteen years. They begin to shed their coats in spring and finish by midsummer. The hair thickens rapidly. By September it is suitable for winter clothes, for which fawn-skins are used. The skins of full-grown animals make tent-covers or rugs.

In winter the herd lives almost entirely on reindeer moss, in summer chiefly on reed-grass and willow-shoots. In late summer and early autumn both moss and grass are necessary to fatten the herd. This is most important, because if a herd does not fatten then it will never fatten and there is a risk of losing the fawns in spring. In the autumn the deer will eat mushrooms, bird-dung round the moulting-places, and even young mice or birds. They are very fond of human urine and are so excited by the smell that they will charge a man who is making water near them. The natives in the camps are very careful in consequence. They use urine to attract the deer when troublesome and it is the most effective means of reviving an exhausted deer on a long journey.

If snow falls late, it is bad for the deer as they cannot walk on ice. The herd must not remain too long in one place because their constant scraping hardens the snow so that they cannot reach the moss. Large herds move every few hours and are therefore leaner than small ones. Hence sufficient space is essential. But summer pastures will stand much more wear and tear than the winter ones. The deep snow of the forests makes it almost impossible for the deer to find food there. Trespassing is a serious offence, as once two herds get entangled it is very difficult to separate them.

Calving goes on from mid-March to the end of May in the herds, three weeks earlier than in the wild state, with the result that many fawns die. During summer the bucks are kept away from the fawns and does. Does often rut in their first year. Hence the rapid increase of a herd. The Chukchee are careful in selecting animals for breeding. Gelded deer or barren does alone are driven. In an average herd the percentage is 12 breeding bucks, 10–15 sledge-deer, and 60 or 70 half-grown fawns. In a large herd there will be 30 bucks to 1,000 does. Wolves are the chief enemies. Hoof-swelling, caused by walking on dry ground, is the principal malady. The first frost cures it, but it often causes a number of deaths. Ticks are troublesome. Far more serious is the scabies that carries off whole herds when it pays its periodical visits.

The herds require careful attention during the breeding-season and still more in summer when they are troubled by insects. Even the women then help to watch them, as the least thing causes a stampede. The herdsmen are often bound to stay two or three nights without sleep. In summer they have to carry everything themselves, including the skins of slaughtered animals, and the weight the less active will carry so as to leave the others free is astounding. But in winter a couple of boys can watch a herd for weeks. Deer are caught with the lasso, and a good lasso is worth a fat buck. The Yukaghir on the Omolon, who only use their animals for riding, keep them in sheds during winter, allowing them two small graylings a day for food. Poor men anxious to own a herd take service with a big breeder. They must work hard, but with luck may own 100 deer after five years.

The least timid animals are chosen for driving and broken in during the first year. With the Lamut deer this is easy, but often very difficult with the Chukchee. One animal is used for a pack-sledge, two for driving. One woman will lead 10 or 15 sledges fastened one behind the other, but a wealthy family may travel with 40–60 in several lines. With good going well-fed deer will do 200 miles in two days, but they need instant rest if tired, and spare animals are therefore usually taken. Dogs have far more endurance.

Thanks to their herds, the Chukchee are much better off than the fish-eating tribes and they are always called in to stave off famine, as well as to supply food in ordinary times—at the Anyui fair, for instance. But the highest famine price for a deer is 16*s*. 8*d*. and their ordinary value is only a cake of brick-tea and a packet of tobacco in fair time. On the Anadir, with its salmon and its wild deer, a shilling is the usual price.

Eskimo

The Eskimo are not a Siberian tribe, but a number of them have crossed over from America and have settled along the west coast of Bering Strait from Cape Dezhneva to Cape Bering, either in villages of their own or in common with the Maritime Chukchee, with whom they are identical in material civilization. They are most numerous near Cape Dezhneva and between Capes Chaplin and Ulyakhpen. They number about 1,600, including those on St. Lawrence Island and the Diomede Islands. Their language is said to be closely connected with that of the Aleuts. Most of those round Cape Chaplin speak a little English. They smoke as much tobacco and drink as much spirits as they can obtain. When their customs and beliefs differ from those of the Maritime Chukchee, they are of American origin. A dying Eskimo is placed in a specially built snow hut or tent, according to the time of year. He is carried in by a back entrance, all signs of which are then removed. He is visited occasionally by relatives, but at the approach of death he is left altogether alone. The Eskimo are a maritime people and hold that their dead live under the sea. The road thither is very difficult and a soul may die again on the way, but once there a man has all he can desire.

Koryaks

Territory.—The Koryak tribe extends from the Stanovoi Mountains to the sea and on the west side of Kamchatka as far south as lat. 55° N. The north-west boundary of their habitat is now the River Varkhalan; they used to extend along the west shore of the Sea of Okhotsk.

Name of tribe.—The name 'Koryak' is not used by themselves, but probably derived from neighbouring tribes. Its derivation is quite uncertain.

Racial affinities and language.—They seem to be closely related by race and language with the Chukchee, but their language is not reduced to writing. There are four main dialects of it spoken by (1) the Koryaks of north Kamchatka, (2) the Reindeer Koryaks of Kamenskoe, &c., (3) the Alutor Koryaks, (4) the Kereks in the NE. The main division of the Koryaks is into Reindeer and Maritime Koryaks ; the manner of life of these two branches of the race has made them develop on wholly different lines ; there is little intermarriage between them because of the difference of their mode of housekeeping ; the Reindeer Koryaks intermarry with the Chukchee, the Maritime Koryaks with the Kamchadals. The Reindeer Koryaks have advanced less far in civilization, but they are generally given a better character. The Reindeer Koryaks are mainly in Gizhiga and Petropavlovsk ; there are few in Anadir, none in Okhotsk.

Numbers.—In the census of 1897 the population was distributed as follows :

Gizhiga . .	4,434	(2,389 Reindeer :	2,045 Maritime)
Petropavlovsk	2,675	(1,284 Reindeer :	1,391 Maritime)
Okhotsk . .	244	(0 Reindeer :	244 Maritime)
Anadir . .	177	(75 Reindeer :	102 Maritime)
Total . .	7,530		

The population increases in the intervals between epidemics and famines, but the Koryaks have suffered from many scourges : syphilis (called the Yukaghir or Chuvanzi disease, because of the route by which it came to them from Russia), two forms of arctic hysteria, small-pox, and measles, the spread of which was attributed by them to their shortage of professional *shamans*.

Relations with Russians.—The Russians first came into contact with them about 1640. Until 1712 the Koryak refused to recognize Russian sovereignty, but not until 1764 did their opposition cease. From 1649, when the fortress of Anadirsk was built, the Cossacks tried to exact tribute from them. But wars have now ended for them, even with the Chukchee, their secular foe. The relations between the Russians and Koryaks are not altogether happy ; the Koryaks resent the Cossacks' demand for transport free of cost, and see in every traveller an official, and so an object of suspicion. They like better the Americans, who practise 'contraband' hunting of sea-animals ; from them they suffer no harsh exactions, and receive supplies much more cheaply than from the Russians. They also appreciate their alcoholic liquors. Few Maritime Koryaks and no Reindeer Koryaks have learnt Russian.

Social institutions.—The Russian Government have divided them into clans, but these were territorial designations and have become misleading owing to migrations. Their own social unit is the family, though families related by marriage have a tendency to draw together ; there are even cases of fraternizing with unrelated families ; members of such alliances were formerly bound to help one another in war, but the absence of war has abolished this aspect of the alliance.

Physical appearance and characteristics.—The Koryaks are described by Jochelson as below average height. They are well developed, have broad shoulders and good muscles. Their hair is usually black (78 per cent. of the men, and 53 per cent. of the women have black hair), bald heads are rare among them. Their eyes are narrow, but not peculiarly Mongolian ; their nose is of moderate width ; they have little hair on the face ; their skin is bronze coloured. Their speech is slow, and they talk in a lazy manner unless they are excited. Travellers give very diverse accounts of their character, but their marked characteristics seem to be obstinacy, austerity, and dauntlessness. They are said to be hard to deal with, unless their customs are understood ; if displeased, they are churlish, rude, and quarrelsome ; if in good humour, they are friendly and jocose. They are truthful and straightforward and do not flatter. They are hospitable and treat their families and animals with kindness.

Art.—They have highly-developed artistic skill, and make carvings in wood, ivory, whalebone, and horn. Among their arts are basketwork decoration and rugs made of reindeer-skin, with ornamental patterns of the black and white fur of the young reindeer.

Occupations.—Their main occupations are fishing, hunting, and reindeer-breeding. Fishing takes place only during the summer months. The fishing implements are of a primitive kind, little affected by the Russians. They are as yet unfamiliar with seine-nets. They use nettle-fibre, which they spin in a primitive and imperfect manner. They use skin-boats constructed like those of the Eskimo; a large boat is nearly 30 ft. long with a maximum width of about 8 ft. between the gunwales. It is covered usually with skins of the thong-seal, the use of which is spreading to other parts where the walrus is disappearing. The Koryaks steer with an oar; they are not really good seamen, though better than the Kamchadals. They also use *kayaks* (boats for one man) and, in northern Kamchatka, dug-outs. Hunting for sea-mammals is of great importance; they hunt for ground-seal and ringed-seal throughout the year, except in the winter months; the principal period for thong-seal is September and October. Their chief weapon is the harpoon, but they also use the mallet for stunning those creatures that have fallen asleep on shore. During the fishing season the Koryaks are too busy to trouble about the seals. The whale industry is long dead: the Koryaks do not go far enough out to sea, but American whalers occasionally bring them dead whales, from which the skin, blubber, and whalebone have been removed. The only animals killed by the Koryaks on land for the sake of food are wild reindeer and big-horns. There are but few of the former; the latter are hunted principally in autumn. They hunt animals chiefly for their fur; the bear (which also furnishes food) is hunted four times in the year, (1) in summer, when it goes fishing; (2) in autumn, when it hunts berries; (3) in winter in its lair; and (4) in spring in self-defence. Hunters among the Maritime Koryaks train dogs, which do not drag sledges, to attack bears. Foxes, especially red foxes, are clubbed, trapped, shot, and poisoned. There are some grey wolves in the tundra. The sable is now rare ; so are the ermine, otter, and glutton.

Reindeer-breeding is still in a primitive stage (see pp. 103–106). It may not be more than a thousand years old. Left to themselves, the reindeer readily return to the wild state. However, the Koryaks will domesticate wild reindeer. Reindeer-breeding necessitates a wandering life, as the herds in search of their food paw up all the snow. The use of dogs for driving is probably very old. Dog-breeding necessitates settled habits, as it requires large stocks of animal food for the winter. The main food of the Siberian dogs consists of fish. The Maritime Koryaks (as also the Yukaghir) build roomy sheds at the side of their houses for their dogs. When spring comes and the sledge is no longer employed, the dogs are given no food, but have to hunt for it. They are very fierce while driven. Should a team of dogs meet harnessed reindeer they will, unless prevented, inevitably tear them in pieces, and a meeting of two dog-teams will lead to a serious encounter, if not forcibly prevented. The average number of dogs possessed in one household among Maritime Koryaks is ten.

Dwellings and furniture.—The habitation of the Reindeer Koryaks is an outer tent with an inner tent for sleeping ; the Koryak tent usually has three or four inner sleeping-tents (*polags*) of small dimensions (6 ft. square and 4 or 5 ft. high) partitioned off with light poles and skin curtains. A camp seldom contains more than three tents. They have four main removals in the year : (1) in October they put up their tents in the river valleys under the protection of high banks among poplar and aspen groves ; (2) in spring, at the end of March, before the fawning period begins, they descend to the open tundras on the lower courses of rivers ; (3) in July they ascend the mountains to be near the river sources ; (4) in autumn, at the time of the fawn-festival, they return from the ridges to the tundras and river-valleys. The

Maritime Koryaks have their dwellings underground, or half underground: one type is described as like an hour-glass in shape; these are permanent buildings of wood, varying in size; they used to be more spacious than now. Among the Kereks as many as twenty-five persons often live in one house. During the winter the lower entrance is closed, and the house is approached by a ladder, or rather a log of wood with holes for the feet, inconveniently small for Europeans. The descent into the interior is disagreeable when there is a smoky fire. In the summer-time the ladder is removed. They import metal kettles, prizing especially copper; for water they use skin or wooden buckets. The atmosphere inside the huts is such that the Koryaks usually sleep naked; sometimes their clothes are put outside for the parasites to freeze off them. The fire-drill is only used ceremonially.

Clothing, food.—They dress in deer-skin, their costume consisting of a *kotlanka* (or frock), trousers, boots, and leggings. In summer their clothes are of dressed skins, in winter of skins with the hair remaining. They are passionately fond of tobacco, which they chew, but rarely smoke. A favourite intoxicant is made of fly-agaric, a kind of fungus, but women never take it; it is a poison, which if taken in very great quantities will kill; it is used by *shamans* to produce an ecstatic state. Brandy, though forbidden, finds its way among them: it is popular, especially with older people. The ordinary food is fish, reindeer meat, dried salmon, and seal's blubber with rancid oil.

Birth, marriage, and death.—The mortality of infants up to one year is enormous, and the number is increased by putting to death any child whose mother dies during or soon after confinement, as artificial feeding is impossible.

The penalties for unchastity are very severe, and illegitimacy is almost unknown. Polygamy is rare; the 'elder' in the settlement is often polygamous, but most cases of polygamy are due to the observance of the levirate law, by which a man has to marry his deceased brother's wife or owing to the barrenness of his first wife. The future bridegroom has to serve for his bride ; the period of service may be anything from six months to three years. If a man does not please his future father-in-law he can be sent away after many years of service without any reward. Money cannot be substituted for service. The preliminaries of marriage are arranged by the 'matchmaker' (asking one). The actual ceremony is by seizure.

When a Reindeer Koryak dies his body is dissected to find the probable cause of death, and the Maritime Koryaks stab the dead man in order that the child in whom his body is reincarnated may not die of the same illness. The dead are burnt, except by the Kereks, who let down their dead in funeral attire into the sea. Parenticide is now abandoned ; even the tradition has disappeared in some places; but it seems to have been a general practice, in order to spare the sick and aged unnecessary suffering. Now relatives take good care of a dying man.

Other customs.—Only clothing and ornaments are personal property among the Koryaks. The wooden 'guardians', household appurtenances, house, nets, and skin-boats are family property. The reindeer are the property of all the members of the family, but the movements of the herd are directed by the father. The Koryak can count better than the Chukchee. He has two bases of computation, 5 and 20, and in counting uses both hands and feet.

Religion.—The Maritime Koryaks have adopted Christianity and renounced many of their superstitions ; the Reindeer Koryaks retain much of their primitive religion, as do also the Maritime Koryaks of Penzhina Bay and north of Alutorski Cape. However, the combined influence of traders and Cossacks has made them abandon a good deal of their religion.

It was among the Koryaks that the *shamans* were first affected by Christianity. In the Koryak houses are wooden images of 'guardians' ; they receive homage as containing a vital principle in them and having had incantations pronounced over them.

The chief religious festivals among the Koryaks are :

(1) Among the Maritime Koryaks :
  (*a*) Whale festival.
  (*b*) Putting away the whale-boat for winter.
  (*c*) Launching the skin-boat.
  (*d*) Wearing of masks.

(2) Among the Reindeer Koryaks :
  (*a*) Ceremony on the return of the herd from summer pastures.
  (*b*) Fawn festival.
  (*c*) Reindeer races.

(3) Ceremonies common to both :
  (*a*) Bear festival.
  (*b*) Wolf festival.
  (*c*) Ceremonies in connexion with fox-hunting.

The reindeer races are religious, in honour of the One on High, while dog-races and foot-races are secular. Every owner of a large reindeer herd arranges races once a year, usually about the close of winter. Sometimes the host sacrifices the reindeer that he has been racing.

Yukaghir

Territory.—The Yukaghir originally extended from the Lena to the Anadir and from the Verkhoyansk Range to the Arctic Ocean ; now they are principally to be found above Verkhne-Kolimsk, along the valleys of the Yasachnaya and Korkodon, and in the region of Alaseiskoe. They are a very ancient tribe, who have been gradually pushed northwards. They were once very numerous, for tradition says that the northern lights were the reflection of their innumerable camp-fires.

Name of tribe.—The word *Yukaghir* is not used by themselves : it seems to be a Tungus word, judging by its termination, and probably means the 'distant ones'. Sauer says that they call themselves Andon Domni, which is probably an incorrect rendering of *Odud omni* the 'people'.

Racial affinities and language.—It is difficult to trace

strong racial affinities for them ; most of them now speak the Tungus language, but there are survivals of their own tongue, which seems to have been highly inflected, and very rich in suffixes and case-endings : it has two dialects, one spoken by Yukaghir and Lamuts who live with them on the Rivers Korkodon and Yasachnaya, the other spoken by the Yukaghir and Yukaghirized Tungus on the tundra between the Rivers Kolima and Indigirka.

Numbers.—The Yukaghir are dying out : their marriages are mostly sterile, and they are a sickly breed. The latest figures of their numbers give them only 754, of whom 388 are males. With them must be included the Chuvanzi, a branch of the Yukaghir who live round Markovo, and who number 453 (236 males), but they have either become Russianized or have fallen much under the influence of Chukchee or Koryaks.

Divisions.—The Anaul formed a division of the Yukaghir living on the Anadir ; they were fishermen and had no reindeer ; they have partly died out and partly become Russianized. The tribal name Odul has been adopted by the Yukaghirized Tungus of the tundra ; the Tungusized Yukaghir call themselves Dutki. There has been much intermarriage with Tungus and Lamut, so that the regular type of Yukaghir has largely disappeared. Like the Chukchee and Koryaks they can be divided into Reindeer and Maritime tribes ; they have also been classified from the names of the rivers along which they lived (viz. Alaseya, Omolon, Kolima, Kongina, Korkodon).

Relations with Russians. Social institutions.—At the time of the Russian conquest they had a well-organized clan system, but it is now much in decay. The only tribal unity that they seem to recognize is that they do not make war among themselves ; no traditions survive among them of a common tribal ancestor. Such clan system as they had was disregarded by the Russians, who have composed clans, which are little more than associations for paying tribute. The Russian law allows the natives to settle their own affairs (with the exception of capital offences, such as murder and mutilation) according to the customs of the people concerned. The elder, who under the Russian system replaces the 'old man', is authorized to punish the clansmen with imprisonment and even physical chastisement. Severity, however, is not often required among the law-abiding and timid Yukaghir. Under the native system the prominent personages in the clan life were the 'old man', the *shaman*, the 'strong man', and the first hunter: the last two offices may be combined in the same personage, and the last is the only one whose duties have not fallen at all into desuetude. There used to be a class of captive slaves called *po* (hired labourers were called *nicil*); among these women had a better position than men.

Physical appearance and characteristics.—The Yukaghir are of short stature; on the average they are the shortest people in north-east Siberia; the men's waists are small, and they have slender and supple figures, moving and dancing gracefully. The women have stout waists, and as a rule short clumsy figures; but there are no really stout figures among either sex. Their children look very weak and sickly, and their young men effeminate. The hair of the Yukaghir is usually dark brown; the hair on his face is scanty; the eyes are dark-brown and more widely open than those of Mongol peoples; the complexion is either brown as the Chukchee's and Koryak's, or it is yellow as the Tungus'. They are the most timid tribe in Siberia, and will submit to any treatment te avoid an oath or curse. They are hospitable to a fault, a fact which is known by their neighbours the Yakuts, who make protracted stays among them and eat up their fish. Though mild and kindly, they do not readily forgive an offence, but their fear of Russian administration is such that they do not often commit murder; for the same reason they are accustomed to render services to the Russians without any remuneration. A desire to imitate the Russians has led them to wash, and soap is popular among them; at the same time they regard lice on the person as a sign of good health. They are extraordinarily honest and truthful, and will spare no effort to pay off the debts incurred by themselves or inherited. Despite irregularities in their lives, they are bashful and modest in speech.

Occupations.—Their chief occupations are hunting and fishing. They hunt the squirrel, glutton, and fox, in order to obtain in their place, tea, sugar, and other requirements. The rifle has taken the place of the bow. They capture reindeer while swimming, having discovered the place where they will come down the river when driven by mosquitoes. The hunters kill reindeer for the entire group that accompanies them during the period of the chase; for fear of the evil eye they give a portion of their booty to strangers. Their only domestic animals are dogs and reindeer; they do not breed horses or cattle, but the Yasachnaya Yukaghir hire horses of the Yakuts for the squirrel-hunting season. They use hemp for fishing-nets, and horsehair has replaced the flexible willow-branches that they previously used. They have such wide-meshed nets that, as they say themselves, 'a bear could get through.' But they have other means of catching fish. A bad fishing season and a bad reindeer year lead them almost to starvation. They say when the fishing is bad: 'there is an old man in Verkhne-Kolimsk, whose heart is harder than Russian iron, and he won't let the salmon out of his cave.'

Dwellings and food.—They live during summer in conical tents (*urus*) made of thin poles, and during winter in small houses made of hewn logs. They are more particular than the Koryaks or Yakuts about their food, and will not eat rancid meat. They are great smokers. Such funds as are over from the purchase of tobacco are used for buying brandy, but they will not drink alone. They share their pleasure with the whole family, including infants in arms.

Birth, marriage, and death.—New-born children used to be killed if the mother died in childbirth. Sterility was regarded as a punishment sent by dead relations, and the *shaman* would be resorted to in order that such resentment might be modified.

Before marriage, chastity is not expected of girls, but

indiscriminate bestowal of their favours is disapproved. Marriage is endogamic, but there are strict laws prohibiting marriage between near relatives. A man serves three years for a bride, and if he is then rejected has no compensation. Polygamy is practised : a man will sometimes spend part of the year in the house of one father-in-law, and the rest in that of another. The Tungus and Yukaghir have to some extent borrowed one another's marriage customs.

The dead used formerly to be placed on platforms which were raised on poles. In the Kolima district it was a custom to distribute the flesh and bones among the relatives of the deceased : these were dried and put in leather bags and then worn as amulets, called ' grandfathers '.

Religion.—A nominal Christianity has not affected the Yukaghir much. Shamanism has a much greater hold upon him. Even the Christian Yukaghir has no Church ceremony till a year or more after his marriage.

Kamchadals

Territory.—The name may be applied either strictly to the principal tribe who inhabit the peninsula of Kamchatka, or more vaguely to some wandering tribes north of the peninsula. There are some tribes, too, like the Palanzi, who live in the ostrog north of Tigilski, and the Olyutorski, who live along the Pacific behind the cape which bears their name, who have close racial affinities with them and the Ukinzi between Cape Ozerni and the River Timlata.

Name.—The name Kamchadal is given them by the Russians : they call themselves Itelmen, and are called Konchalo by the Koryaks.

Racial affinities and language.—The race is mainly a half-breed between the aborigines and Siberian emigrants or escaped convicts : the pure Kamchadals are very rare. They have many attributes, especially in costume and customs, in common with the Mongols, but share more with the dwellers in north-east Siberia and north-west America. They are found in the Kuril Islands, especially in Shumshu, the northernmost of the group. The language cannot be assigned to any known group: it is very guttural, and has many inflexions and prefixes. The vocabulary is very poor, there being only one word for the sun and moon. It is most spoken in the south and in the north about Penzhinskoe, where it is purest; but it is disappearing, and most of the tribe speak Russian.

Divisions.—There are three divisions of the Kamchadals: one group occupies the valley of the River Kamchatka, the second the west coast from Bolsheryetsk to Oblukovina, the third the Kurils, where they are found together with the Ainus.

Numbers.—Drink and illnesses have reduced the population. The last figures give 2,805, of whom 1,415 are males, but it is not known precisely which tribes were included in this numeration, and the real Kamchadals are possibly only half that number. They are not a very prolific people: women usually have only four or five children.

Relations with Russians.—Since the suppression of the revolt in 1731 the Kamchadals have been quiet, and they are now largely Europeanized: European have taken the place of native dances; the native costume is discarded for something like that of a Russian peasant; they have also largely given up their extreme fondness for dirt.

Physical appearance and characteristics.—The true Kamchadal in general is below the common height; his figure is round and squat, his eyes small and sunken, his cheekbones prominent, his nose flat, his hair black, his beard scanty, his complexion brown or yellow. He is mild-tempered and honest, an easy prey to traders who deceive him, apt to get drunk, lazy, and apathetic, with no thought for the future, but careless and indifferent. They used to be a warlike and revengeful people, but they are now more remarkable for their readiness to oblige and their hospitality. Lansdell attributes to them a custom of tactfully relieving themselves of a guest whose protracted stay threatens to exhaust their stock by serving him a dish called *tolkootha*—the dish is found among

Tungus tribes also—which consists of a mixture of meat, fish, and vegetables. The guest takes the hint and departs the next day.

Occupations.—Their chief occupations are fishing, especially for salmon, and hunting. The efforts of the Government to introduce cattle-breeding have failed; agriculture does not flourish, as corn will not ripen (except round Klyuchevskoe); gardening prospers better, as roots will grow. Their method for catching salmon, as described by Demidoff, is to fix rows of inclined birch-stakes across a river from one bank to the other with only a narrow aperture on one side for canoes. Attached to these poles a little below the surface of the water and a few yards apart, are set two or three long wicker baskets according to the width of the river. The fish, which come up, are unable to proceed on account of the stakes: they then make their way through the gaps leading into these baskets, out of which inward-turned spikes prevent them from escaping. When the natives go to collect their catch, they lift part of the basket out of the water and secure the fish with iron-edged gaffs through a small door at the top. In this way they manage to take 2,000 fish in a day. They seldom use seines, but almost always common nets, made of packthread purchased from the Russians, or of nettle-fibre: they also use harpoons. They hunt reindeer, big-horns, foxes, otters, beavers, hares, and sables: special methods have been adopted to protect the last, which would otherwise become extinct. They trap bears, and show great patience when they lie in ambush for them. Their chase is attended with certain superstitions: they abstain from washing themselves, they are careful not to pronounce the name of an animal that they hunt for fear of ill-luck, and not to make the sign of the cross. They invoke their god Kutkhu and sacrifice in his honour the first animal that they catch. They are indefatigable walkers, but are also experts in driving sledges and training sledge-dogs. When in a team, the most intelligent dog is selected as leader; the others are harnessed two and two behind. A cry of *tag-tag* makes them turn to the right, a cry of *kougha* sends them to the left. The harness is of leather: it is passed over the dog's breast and is joined to the sledge by a strap 3 ft. long in the manner of a trace. If the driver strikes the ice with his stick (*oshtol*) they go to the left; if he strikes the side of the sledge they go to the right; if he places the stick in front of the sledge, they stop. The dog-sledge is practically their only means of communication and horses are very rare.

Dwellings.—Like many other Siberian tribes they live in different kinds of huts during summer and winter. The former (*balagans*) are erected on posts about 12 or 14 ft. high; their conical roof is covered with a kind of thatch made of bark; the cooking is performed in the middle of the room where they all eat and sleep together; there are no windows and the doors are so low that they scarcely admit the light. The staircase is merely a beam jagged in an irregular manner; if it is turned with the steps, or notches, inward, it is a sign that the residents are not at home. One advantage of the height of the house is that they can dry their fish out of the reach of the dogs. Their winter houses (*izbas*) are of wood: they are made of trees placed horizontally with the interstices filled with clay; the interior usually has two rooms, which can be warmed, as in Russian inns and small houses, by a stove set between them. The insides are tidy and often decorated. Windows are made of skins of salmon or bladders of various animals.

Clothing, food.—Lesseps in 1790 describes their costume as an outer garment (*parka*) made of skins of deer or other animals, tanned on one side, and long breeches of similar leather; next the skin is worn a very short and tight shirt of nankin or cotton, the woman's being of silk. They wore fur caps. A recent traveller, Demidoff, says that now their costume resembles that of a Russian peasant—a blue cotton shirt under an old brown jacket, broad trousers tucked into topboots, and a military cap. Their boots are made of rein-deer hide, the soles being stitched on to seals' throat-skins round the calves. In summer they wear boots of goats' or dogs' skins tanned. Their principal food is dried fish; some fish they allow to become putrid in a hole and then eat them.

Birth, marriage, and death.—Births take place in public, with relatives and neighbours gathered round. Infanticide is practised, women giving their undesired offspring alive to the dogs; if twins are born, one of the pair must be killed; so must a child born during a storm unless incantation can remove the evil that would ensue.

A man's bride is usually selected from the next village, not from his own; he serves for her, but is given compensation if he fails in his suit. He has to capture his bride as among the Koryaks, but the ceremony is more of a reality. Marriage is only forbidden between parents and children. Virginity is not required in a bride. Divorce is easy.

The dead are eaten by dogs; children are buried in hollow tree-trunks.

Religion and superstition.—Their chief god is Kutkhu, the supreme being; his wife with them is called Kakee, his son Trel-Kutan, his daughter Shi-Shakels. Their mythology is crude and obscene. Volcanoes and hot springs are the abode of evil spirits (*Kamuli*). Sacrifices are not made to the gods, but to the many spirits good and bad with which they people heaven and earth, the greatest of whom is Pikhlyash. There is little professional Shamanism among them; every old woman and woman in man's clothes is counted as a witch. There was a class among them called *Koekchuk*, who were treated as women; it is possible that they were captive slaves who were purposely rendered effeminate to make them less dangerous, and who therefore were made to share the woman's life. Certain trades were regarded as unmanly; if a man became a tailor or shoemaker, he was regarded as a *koekchuk*.

Gilyaks

Territory.—The Gilyaks extend along the coast of the mainland on either side of the mouth of the Amur, from

Tugurski Bay on the north-west to the Mamia Rinzo Strait on the south-east, and they also occupy the northern part of Sakhalin down to lat. 50° 10′ N. on the west shore and to about lat. 51° N. on the east shore, the southernmost settlements being respectively Porokolan and Chamr-vo.

Name.—The name by which they know themselves is Nibch (= the men), but the Russians have called them Gilyaks, a modification probably of the Chinese designation for the Kilor or Kiler.

Racial affinities and language.—They present one of the greatest ethnological problems in all Asia. They have been variously claimed as a branch of the Ainu (the race that inhabits south Sakhalin and Yezo), of the Tungus, and of the Tartars—an error which is repeated in the name 'Gulf of Tartary' applied to the sea between Sakhalin and the continent. Some of their characteristics have been regarded as Caucasian, and it has been supposed that there was a large infusion of the blood of Russian adventurers from the seventeenth century. Were it not for their language they might be regarded, so far as their physiognomy and bone structure goes, as a branch of the Tungus; in many of their customs they approximate to their neighbours, such as the Olcha and Goldi, but their speech is quite distinct and cannot be classified. It is an isolated tongue like that of the Koryaks and Yukaghir, and even one unacquainted with the language can on the most casual training distinguish it from any Tungus speech. It is harsh and full of consonants; sibilant, nasal, and guttural sounds prevail. It has many words borrowed from other languages, but apart from its vocabulary it bears no close resemblance to any Mongol language. As far as language goes therefore, the Gilyaks must be classed among the Palaeo-Siberians, but it is possible that they are a people, like the Normans and Bulgarians, who have learnt the language of the conquered, and that a great infusion of Mongol blood in the past has profoundly modified the real type.

Divisions.—There are, however, three types of Gilyak physiognomy, one of which approximates to the Ainu, another to the Mongol or Tungus, while the third is typically Gilyak. There is also a geographical distinction between those of the mainland and the two tribes which live on Sakhalin, Smerenkur on the west, and Tro on the east.

Numbers.—Their numbers are now 4,649, of whom 2,556 are males. They are dying out. Their women have few children. Six is considered a large family. Because the population is dwindling, clans have had sometimes to adopt individuals or whole groups.

Relations with Russians.—The Gilyaks have been less spoilt by civilization than many tribes. They have been known to the Russians since the seventeenth century. For years they succeeded in keeping the Chinese traders out of their land, and they have not become demoralized by intercourse with Chinese and Japanese. But the acquisition of Gilyak land by Russian settlers has not had a good effect on them

Social institutions.—They have a highly developed clan organization, with its common fire, common enemies, common obligations of revenge, and common *thusind.* The last is the name for the compensation exacted in place of blood-revenge and in recompense for certain crimes.

Physical appearance and characteristics.—The typical Gilyak is below medium height, of stronger build than his Ainu or Tungus neighbours ; he has a well-developed chest, moderately broad shoulders, short neck and fairly big head, but small hands and feet. There is no superfluity of fat. The complexion is brown, the hair is less abundant than that of the Ainu, but grows longer on the head and more freely on the face than among the Mongols and Tungus. The eyes are small and sparkle with a dull light, the lips have been called 'voluptuous', the nose is rather flat, the cheek-bones prominent, and the eyebrows are bushy. They do not shave the head, but wear the hair tied up in a thick tail or in tresses.

They are an energetic people and temperate in the use of spirits. They prize their tribal and individual liberty. Their principal faults seem to be avarice and covetousness, and the islanders have had a reputation for theft. Their aloofness from civilization has made them less ready than other tribes to adopt habits of cleanliness.

Occupations.—The men's occupations are mainly hunting, fishing (for sturgeon, salmon, &c.), and trading. They are adventurous in hunting the bear, but their courage is not equal to entering the water, and, though fishermen, the Gilyaks cannot as a rule swim. They are expert in the use of bow and arrows, and are good mountaineers. In rowing the Gilyaks scull, but pull the oars alternately. In fishing they use in some parts gill-nets and seines, and in others scoop-nets; for their nets they use the stalks of the nettle in place of flax.

The man's work takes him much from home; a great deal of work at home is done by women, who occupy a low menial position. Slaves are bought from the Ainus and Goldi. They do not, however, hold or sell their own people as slaves. There are not many slaves, as a female slave costs more than a wife. The slaves have no rights at all; they have to perform the heaviest housework, hewing wood and drawing water.

Dwellings.—Their *yurta* is a wooden house, of which the interior is often divided into an ante-room and an inner room which is inhabited. In the centre of the room burns the fire with a hole in the roof above it for the smoke to escape. The windows are of fish-skin. The walls and the floors are made of trunks of trees, the interstices being filled up with birch bark or leaves, and the roof being covered with birch bark. They used to domesticate ermine to kill the rats and mice, and the Manchus supplied them with cats at a high price, but always castrated so as to keep the monopoly in their hands. Their winter dwellings are in small groups of from two or three to a dozen. In 39 villages Collins counted 140 houses.

Clothing, food.—In winter they dress in dog-skins or the skins of the fox or wolf. In summer they wear fish-skin, which has given them the name of *Yupitatse* ('fish-skin people') among the Chinese. They often wear blouses of Chinese pattern. Their boots are of seal-skin or sometimes cotton. Men and women dress much alike, but the woman's garb is distinguished by metal disks round the bottom of their blouses. The skins of salmon are stripped off very dexterously; they are then beaten with a mallet, so as to remove the scales and render them supple. This gives them waterproof clothing. They live almost entirely on fish. But occasionally they eat animals killed in the chase and even dogs, as do the Ainu and American Indians. The fish is prepared with herbs, roots, and train oil; sometimes they procure a little millet or rice from the Manchu and Japanese in exchange for furs. They do not cultivate the ground themselves. The use of bread, tea, salt, and sugar they have learnt from the Russians. Bread is regarded as a very great delicacy.

Marriage and death.—Chastity is not demanded in a bride. Marriage is exogamic. There seems to be no settled form of marriage, and there is a certain amount of polygamy. The price of a bride is the chief bar to polygamy, but it is on the other hand a great incentive to industry.

Death is supposed to result from the action of evil spirits. Burial rites are of an imposing character. The body is first burnt on a funeral pyre, and a small wooden house is erected over the ashes after they have been carefully collected. The deceased's favourite dog, which has been previously fattened, is killed over the grave.

Other customs: the tiger and the bear.—If a man has been killed by a tiger, superstition forbids any ceremonies at the burial of his body. The tiger is much feared, and his appearance is supposed to portend evil. Their most characteristic ceremonies are connected with the bear, who is called *Mafa* (Chief Elder). There is a bear cage near every village, and in January of each year there is a solemn bear-sacrifice, and at other times a procession in which the bear takes a less exacting part. A bear must not be killed by surprise, for they fear his posthumous anger; they always catch or kill him in fair fight. It is regarded as a happy death to be killed by a bear.

Religion and superstition.—The highest benevolent god of the Gilyaks is Ytsigy according to Schrenk, but according to Sternberg they call him Kurn, by which name also they call the Universe. The 'owner' of the mountain is called Pal; the 'owner' of the sea is Tol. Every natural object has a life of its own and an 'owner'. In their belief also Sakhalin conceals an immense deity. There are besides a great number of spirits, good and bad. Such is their belief in the gods' ordering of the world that they will not save a man from drowning for fear of thwarting the will of a heavenly power. Ancestor-worship permeates their religion, and supports their clan-system. They have many taboos, and among the most rigid of their restrictions is the custom by which no one but a clansman may remove fire from a *yurta.* This is so strict that a stranger must always be careful to finish a pipe before he leaves a house. The Gilyak seem to have been indebted to the Goldi for much in their ritual, customs, ideas, and art.

Among the Gilyaks of Sakhalin are a number of isolated settlements of Oroke (a Tungus tribe). The southern half of the island is inhabited by the Ainus (a Palaeo-Siberian tribe), but their habitat is entirely included in the part that belongs to Japan.

Ostyaks of Yenisei

Territory.—This tribe lives along the course of the middle Yenisei and its tributaries between Miroyedikha, near the mouth of the Lower Tunguska, and Yeniseisk. They are most numerous about Sumarokovo. They were probably once more widely extended.

Name of the tribe.—Though called Ostyak, they have nothing to do ethnologically with the Ostyaks of the Ob basin. They do not even, it appears, belong to the Ural-Altaic stock. They call themselves Tindigyet. Kanacket and Din (people).

Racial affinities and language.—Their origin presents a difficult problem, which seems to defy solution. They are thought to be a remnant of the primitive people who were the original inhabitants of Siberia, the centre of whose civilization was further south. Their language is unlike any other known tongue. Most of the river-names in the neighbourhood of the River Tom belong to it.

Numbers.—They are now not as many as 1,000 in number, and they are diminishing. Their principal foes have been syphilis and alcohol, both of which have had deplorable effects in reducing their numbers.

Relations with Russians.—They have become deeply in debt to Russian traders, sometimes owing as much as 500 roubles. They do not intermarry or have sexual intercourse with Russians.

Physical appearance and characteristics.—Their faces are of two types: one is short and broad with heavy cheek-bones, typically brachycephalic, the other approximates more to the Aryan type, and is longer. Their hair, though dark, is finer and lighter than that of any other inhabitants of the Yenisei valley. They have not the chief Mongolian characteristics: e.g. their eyes are not oblique.

Occupations.—Their principal occupations are hunting, fishing, and reindeer-breeding, but they have only taken to the last in recent times. They hunt elk and squirrels. For fishing they use canoes, not made of birch bark, but hollowed out of the trunks of trees. A bad season in fishing or squirrel-hunting impoverishes them badly, and epidemics of anthrax among their herds have brought about the decay of the tribe. But despite their poverty they seem a contented people. During the fishing season they live in birch-bark tents along the river banks.

Costume.—The special feature to notice is the men's habit of wearing a handkerchief round the head, as is so often done by women in other lands.

Religion.—Nominally members of the Russian Orthodox Church, they have remained comparatively faithful to their old traditions.

Aleuts

The Aleuts are found in the Aleutian Islands, which now belong to America, but a few of them are found in the Commander Islands, off the coast of Kamchatka. They are of low stature, but well shaped; they have dark faces, black eyes, long black hair and short necks. They are nominally Christians, but seem to have assimilated more of the bad habits of professing Christians, than of their doctrines.

II. The Neo-Siberian Tribes

(i) Finno-Ugrian Tribes

Lapps

Territory.—The Lapps in Russia occupy the whole of the interior of the Kola Peninsula, and some live on the coast in the Ponoi district. They extend west into Norway, Finland, and Sweden.

Name.—Their own name for themselves is Same, and for their country Sameland.

Racial affinities and language.—The Lapps are a branch of the Finno-Ugrian tribe. Their language in some respects resembles the Mordvinian speech, but the general system of conjugation and declension is like Finnish, from which tongue, however, it differs phonetically by its great number of diphthongs and consonants.

Numbers.—In 1897 there were 2,040 Lapps in Russia, of whom 1,590 were in the Kola Lapp district and 450 in the Ponoi district.

Divisions.—Among the divisions in which the Lapps are grouped are the Lyavozersk Lapps and Ponoi Lapps. The former, numbering 349, are in four villages, and have been little influenced by Russian manners; the latter, numbering 450, are in six villages, and have been much influenced by Russian customs. The two groups speak a different dialect from one another and are mutually very suspicious. They are not divided, as in Finland, into Fisher and Reindeer

Lapps, but each family as a rule practises both occupations, tending reindeer in winter and fishing in summer.

Relations with Russians.—Different parts of the country have been differently affected ; the most Russified have been the dwellers in the Ponoi district, and along the coast generally there has been much interbreeding with Russians and Norwegians. The race is losing its national characteristics and is degenerating. In 1897 the population of the whole of Kola Peninsula included 6,020 Russians, 850 Finns, 230 Norwegians, as opposed to 2,040 Lapps.

Physical appearance and characteristics.—The Lapps are the shortest and most brachycephalic race in Europe. They are a dwarfish and thick-set people. In complexion they are generally fair, with long shaggy hair, which is usually dark-brown, with a good deal of yellowish hair on the face, the beard being often cut to a point. The eyes are narrow, but set horizontally, the nose is broad, the mouth big, the chin pointed, the cheek-bones high and prominent, but not so prominent as those of the Samoyedes, who altogether have a more Mongolian look. They are cleaner and have more pleasant manners than the Samoyedes ; they are not intellectual, and, unlike the Finnish Lapps, from whom they also differ in dialect and creed, they are unable to read and write.

Occupations.—In spring and summer the Lapp families engage in fishing, especially for salmon, in regions which the custom of each village defines fairly closely. The summer villages are found from 10 miles north of Voroninsk eastward to Paitspahk and the sources of the River Ponoi, and south to the Umpjavr Lake. Many of them migrate to the sea-coast with their reindeer. Lake Imandra, on the old post-road from Kola to Kandalaksha, both in summer and winter, has Lapp villages around it. Reindeer-breeding is their great occupation, but they are very unscientific and unbusinesslike in their attention to it. In summer, when they do not want the deer for travelling purposes, they let them roam in freedom, on the bare heights of the interior, when they become half-wild, breed at will, and sometimes stray away altogether.

In winter they are left near the winter village, feeding on the reindeer 'moss' which they reach through the snow by pawing with their hoofs. There they stay till some time in May, when the new-born calves can be marked. The consumption of the lichen necessitates the removal of the Lapps' villages every 15 or 20 years, but a minor migration takes place each year in April or May, when they pack up the windows and more valuable things, abandon the villages, and go off to their summer quarters. Each family generally possesses from fifty to two hundred reindeer, but real independence is only attained by those who have as many as three hundred. Really rich men have a thousand or more, but such wealth does not make them change their simple life. The boat-sledge in which the reindeer is driven is called *kereoshka*; the Lapps drive without the *harray* or long pole which is customary among other reindeer peoples. Generally speaking their sledges and methods of driving are inferior to those of the Samoyedes. In winter the Lapps do some business in the transport of goods. They trade with Russians and Zirians, who exchange knives, powder, and other objects of barter for their reindeer-skins. They leave all enterprise on the sea itself to the Russians and Karelians.

Dwellings.—The settlements of the Lapps are called *pagosts*. They have summer and winter villages, the former near the sea-coasts and lakes, the latter near the forests, where they herd their deer. The summer-dwelling, called a *viezha*, resembles a Samoyede *chum*, but is not covered with skins, but with branches, tree-bark, and turf. The winter-dwelling, called a *tupa*, is a small, smoky, sod-covered timber hut, some 150 to 200 square ft. in area. These huts are always in groups or villages. A good example of a Lapp settlement is the village of Lyavozerski, on the west of the Lujavr lake and on the south bank of the Varmyok stream. In 1887 it contained 61 males and 63 females, in 19 wooden huts and 7 turf huts. The church was nearly two miles to the west, where the village had originally been, until want of wood and lichen caused its transference.

Clothing and food.—In summer the Lapps dress like Russian peasants, common features being a grey cloth jacket and high woollen cap with a tassel at the top. There is nothing distinctive about the women's dress, which is usually a calico *sarafan*. The Lapp has no strong prejudices in favour of a national costume, and is ready to wear whatever he can get hold of. In winter they wear garments of reindeer skin. The *pesk* is a fur garment with the hair outside, somewhat like the Samoyede *suvik*, but without the mittens being attached to the garment. Instead of a hood they have detachable caps. They wear short fur boots, known as *kadzhi*, and also longer boots made of reindeer-skin, called *yerra*, with the hair removed above the knee.

Throughout the winter they have for food reindeer-flesh and dried and salted fish. At various times they get snow-chickens, water-fowl and their eggs, and berries.

Religion.—They have assimilated Russian religion, just as they have assimilated Russian customs, but it may be questioned whether the adoption is not even more external. They have a name for being very superstitious, and an association of wizardry and vaticination goes with the name of the Lapp.

Karelians

Territory.—The bulk of the people of this name live in the Governments of Olonets and Arkhangel, and there are considerable numbers at Tver and Novgorod. The country which bears their name is part of the district of Kem in the Government of Arkhangel, lying to the north of Pomorland or Pomoria.

Racial affinities and language.—They are closely allied to the Finns, but are a distinct tribe, and show certain differences as the result of having come much more under Russian than Swedish influence. They are first mentioned in the ninth century. The tribe to which they belong overran most of the south-west coast of the White Sea till the fourteenth century, when they spread eastward and occupied the banks of the Dvina. They speak a tongue like, but distinct from, the Finnish of Finland: the spoken tongue is corrupt and has become mixed with a number of Russian words, but among the Karelians farther to the south-west was collected the *Kalevala*, the great Finnish epic.

Relations with Russians.—They have come much under Russian influence, and, unlike the Finns, have never been subject to any other European nation. But owing to their remoteness they have preserved their customs more than their western neighbours. Those of them who live near Russian settlements can usually speak Russian, though they speak their own language among themselves.

Physical appearance and characteristics.—The Karelians resemble Russians: their eyes are usually blue, their hair is brown or ruddy; their forehead is low with hair clipped down over it, level with their eyebrows and hanging down evenly behind. They are slighter in build and better proportioned than the Finns. They show themselves more enterprising, vivacious, and sociable, but they exhibit less perseverance.

Occupations.—Their occupations are very various. Agriculture means a great struggle against the forces of nature. The best and most lucrative employment is the felling, transport and floating of timbers for the saw-mills. River- and lake-fishing provide only a small income, and have nothing but local importance; but they also do sea-fishing in the Gulf of Kandalaksha for salmon, herring, and marine animals. Hunting of game in the forests was profitable, but it has died out since the law in 1892 against trapping. The people require good guns to secure success again. The carrying trade with Finland was more prosperous before the Finns opened a shop in almost every village, a step which greatly reduced their trade with the Karelians.

Houses and furniture.—The Karelian houses are built on a sort of permanent scaffolding: they are reached by ladders. The door is generally on the left, and a corridor divides the rest of the habitation from the store-shed. On the ground are sheep-pens and cattle-sheds. The kitchen utensils are poor, consisting as a rule of nothing but a kettle, a water-tub and a few spoons. Only those who are well-to-do have anything like a *samovar*, but earthenware is imported from Arkhangel.

Clothing and food.—The men wear an outer dress of grey cloth, somewhat like the smock frock of the Little Russians, underclothing of coarse linen, and boots of yellowish leather with leggings attached. Their head-dress is anything that they can find in the way of hats or caps. The women's smock-frock is much like that of the men, and they wear a *sarafan* of striped or printed calico. For footgear they have shoes and for headgear kerchiefs or headbands. In winter long sheep-skin coats are worn. Those who traffic across the border of their district are apt to imitate town fashions.

Their principal food is *ukha* or *tchi* (a soup of fish and vegetables). Most mix their flour with bark and straw. On holidays they eat fish-cakes (*ribniki*): on fast-days their fare is salted mushrooms and edible fungi stewed with turnips and potatoes. When the harvest is good, they brew a sort of country beer, called *braga*, but vodka is not drunk among them. Tea is a luxury of the rich, but is much appreciated when obtained.

Religion.—They belong to the Orthodox Church.

ZIRIANS

Territory.—This important tribe is found in the Governments of Perm, Vologda, and Arkhangel: it inhabits part of the Pechora district of the Arkhangel province, and the whole of the Ust-Sisolsk district and two-thirds of the Yarensk district of the province of Vologda. Its centre is Ust-Izhma. Formerly the Zirians extended further west.

Name.—There are many forms in which their name is spelt: they are known variously as Syryenians, Zyrenians, Sirianians, Zirianians, Zyrians and Zirians. Their own name for themselves, however, is Kami.

Racial affinities and language.—Like the Permyaks and

Votyaks, they are part of the Permian branch of the Finns. Their language was early reduced to writing, but they have no literature nor written memorials. It is sufficiently like the language of the Permyaks for the two peoples to be mutually intelligible. It is freely supplemented by Russian words and has a Samoyedic element in the vocabulary associated with reindeer, for they have derived from the Samoyedes the art of breeding and herding deer. A grammar of their tongue has been published by the celebrated Finn, Castrén.

Numbers.—It is estimated that there are 85,000 altogether in European Russia, and another 1,000 in Asia along the lower Ob. In 1899 they formed 60 per cent. of the population of the Pechora district, distributed as follows: 1,780 in Ust-Kozhvinskoe, 1,730 in Kevda, 5,590 in Krasnoborsk and 12,000 in Mokhcha, the total number being 21,120.

Relations with Russians.—In the ninth century the Finns and Russians were living in amity together in north Russia, the Finns paying tribute. The Zirians have been in constant relations with the Russians, have learnt much from them, and are likely to be absorbed by them. It seems probable that they will lose their language: they sing Russian songs without any idea of their meaning, and have adopted much that is Russian in customs and costume. The most important change is that they are ceasing to be nomadic. They have assimilated the village institutions of the Russians.

Physical appearance and characteristics.—The Zirians are a people of medium stature and robust frames: they are blond and grey eyed: they do not differ greatly in appearance from an ordinary peasant of Arkhangel. The best Zirians are those that dwell about Ust-Izhma, and those who live at Mokhcha. These are vigorous and vivacious, and devote themselves with success to commerce. The other Zirians are apt to be inert and unenterprising. Outwardly the Zirians are very devout, and most of their villages have a well-built church. They are noted for their hospitality even when they are very poor. Their morals are inclined to be easy.

Occupations.—Their chief source of prosperity is reindeer-breeding, an occupation in which they have been engaged for a very long time. The prosperity of the Mokhcha and Izhma Zirians can be seen from the numbers of reindeer which they possess. In the Pechora district in 1896 there were 276,315 head of reindeer. Of these the Samoyedes owned 49,950, the Mokhcha Zirians 194,520, the Russians and other Zirians 35,245. A herd of 500 reindeer is calculated to bring in £50 per annum clear profit, and single Zirians possess as many as 4,000 reindeer each. They employ Samoyedes on wages as labourers and herdsmen. They seldom know themselves how many deer they possess. They slaughter about one-fifth of their stock annually. The Zirians are also engaged in agriculture, hunting, and trade. The cross-Ural trade in the wares of the Samoyedes is very largely in their hands.

Dwellings.—The Zirians live in log-houses (*kerkas*), of much the same style as those of Russian peasants. The house has two rooms, one a rather untidy living-room, with a Russian stove (i. e. a brick stove built into the wall) in it; the second, a reception-room, only used on particular occasions, with a Dutch stove in it. Between the two rooms is an entrance-hall, which opens into a *poviet* or general store-shed. This store-shed is entered by a sloping incline of logs. Often a steam bath-house is found near by, beside a river. The Zirians use the bath-house several times in the week, and in summer plunge straight from the bath-house into the river. Their villages are often of substantial wooden houses. In places like Ust-Izhma and Mokhcha, where there are well-to-do Zirians, there are houses of two stories. In their case the lower floor is like an ordinary *kerka*, as above described, but the upper floor has a sort of drawing-room, with stuccoed ceiling and painted floor covered with reindeer skins.

Clothing and food.—The ordinary winter costume is a *malitsa* made of reindeer skin: this is a huge fur overcoat, in form like a sack, worn with the fur inside, having a high collar and mittens attached to the sleeves. The *suvik* is a larger malitsa worn with the fur outside and having a hood sewn to the collar. The Zirians wear high boots (*pimi*) with the fur outside; in summer they wear no cap, but only a sort of hood.

Their principal articles of diet are meat, fish, and milk: bread and vegetables have only a secondary place. They are slovenly in dressing their meals, never washing meat or fish; the latter they find it difficult to keep, owing to scarcity of salt. They are fond of *vodka*, with deplorable results; they also brew from barley-malt a sort of *kvas* and beer. They appreciate tea, but they mix it with pepper, onions, and aniseed.

Religion.—They were converted to Christianity in the fourteenth century by St. Stephen, who translated the gospels into the Zirian tongue. They are members of the Orthodox Church.

Voguls

Territory.—The Voguls (also known as the Maniza) are found on both sides of the Ural Mountains, but the bulk of them live between the mountains and the Irtish and Ob, extending as far north as the River Sosva: they are most numerous about the River Konda. The European Voguls are found in the Government of Perm, between the headwaters of the Pechora and the Urals. Formerly they extended further south and west.

Racial affinities and language.—They are a branch of the Ugrians, and so a Finnic people: their language is akin to the Hungarian and to the Ostyak speech.

Numbers.—Their numbers were estimated in 1912 as being 7,476, of whom 3,720 were males. Of this total about 2,000 live in Europe and the rest in Siberia. They appear to be decreasing in numbers.

Physical appearance and characteristics.—They are not unlike the Ostyaks, having round broad faces, broad noses, prominent cheek-bones, and black hair, and are small in stature. They are said to be the least sociable of the Siberian aborigines.

Occupations.—Like the Ostyaks, they are mostly hunters and fishers: they have little pasturage, and practically no agriculture. There are certain rules for the preservation of game that they scrupulously observe: they remain only a certain time in one encampment, and no encampment may have more than five yurts in it, and they must be at least ten miles from any other encampment, because the smoke from their dwellings drives away game. Consequently they live in isolated groups, a practice which would either encourage their lack of sociability or account for their reputation as unsociable. They rear reindeer, but have few horses. They trade with Samoyedes, Ostyaks, or Russians, principally in furs, going as far north as Obdorsk for commercial purposes.

Clothing and food.—They usually wear Russian dress. They either make or used to make summer clothing of nettles gathered in September, and woven into garments. Their food is principally fish or reindeer meat. A somewhat liberal interpretation of what was fit for food has become more restricted with the growth of civilization.

Marriage and death.—A hunter may have more than one wife, according to his means, but the union is easily dissolved and the husband often lives alone. When a Vogul dies the body is taken out not through the door, but by a window or specially made hole. The graveyard is usually in a forest, the body being brought there by reindeer, which have to be slaughtered in a particular manner. The body is laid in a boat or in a coffin shaped like a boat. After the funeral there follows a feast.

Religion.—They are nominally Christians, but a good deal of Shamanism survives among them.

Other Finnic Tribes

The other Finnic tribes fall outside the scope of this book. The Permyaks are scarcely to be distinguished from the Zirians, whose language they can understand. They are found in the Government of Perm, and particularly on the River Kama. The Votyaks are a numerous race, about 250,000, found in the south-eastern part of the Government of Vyatka; they are physically weak, and have no striking mental qualities. The Volga Finns, the Mordvinians, number about 1,860,000, and live on the middle Volga about Kazan, Kostroma, and Vyatka, and also in Ufa and Orenburg.

Ugrian Ostyaks

Territory.—The Ostyaks live in the region of the lower Irtish and the lower Ob from its junction with the Irtish to lat. 67° N. Eastward they extend as far as the Tomsk district and the Yenisei. At the time of the Russian Conquest their abodes extended much further than now. Remains of their fortified places, destroyed by the Cossacks in the sixteenth century, are to be seen in several parts of the country, many of them being found in the neighbourhood of Obdorsk.

Name.—They call themselves As Chui or As Yakh (i. e. people of the Ob), and of this the name Ostyak by which they are called is a corruption, unless it is a corruption of the Tartar name for them Uemtäk (= barbarians). The northern Ostyaks call themselves Handocko (= the men).

Racial affinities and language.—They belong to the Finno-Ugrian tribe, and are members of that Ugrian branch of which the Vogul and Magyar are also members. They are closely related to the Vogul, but that did not prevent them from being constantly engaged in warfare with them till both were conquered by the Russians. They all speak one language, but there are three or four leading dialects. The speech of the Ostyaks round Berezov is like Vogul and so different from the language of the Ostyaks round Obdorsk that the dwellers in the two districts cannot understand one another. The Ugrian languages seem to have become separated from the other Finnish tongues before the development of the system of conjugations and declensions. The Hungarian tongue has come much under European influence and is much more highly developed than either Ostyak or Vogul, but it presents certain close affinities. The Ostyak has a difficulty in pronouncing the letter *f*. The purest dialect is said to be at Surgut.

Numbers.—The numbers estimated in 1912 were 17,221, of whom 9,012 were males. The numbers seem to be decreasing; infant mortality is very high, and they are much afflicted by famine.

Divisions.—The Ostyaks are ordinarily divided into Ob Ostyaks and Irtish Ostyaks. The difference is not merely geographical, for the Irtish Ostyaks are superior in development to the rest of their race; they lead a more settled life than the Ob Ostyaks who are mostly nomadic. The 'Ostyaks' of the Yenisei are a wholly different race.

Relations with Russians.—In the fifteenth century the Irtish Ostyaks were much influenced by Syrgan settlers from East Russia, and in the sixteenth and seventeenth centuries they were much influenced by the Tartars. In 1897 certain southern Ostyak villages still spoke Tartar, and Mohammedan influence appears in certain popular customs, for instance, in the avoidance of eating pork, and in the custom of women covering their faces before strangers. Russian civilization has for years now been making great advances, chiefly through the marriage of Ostyak men with Russian women, rather than *vice versa.* Thus many Ostyak villages in the Irtish region speak only Russian, and although old men may speak Ostyak, the young men are not learning the tongue of their fathers. The trade between the Ostyak fishers and the Russian merchants from whom they purchase corn for bread, the use of which has become greatly extended, has helped the Russification of the Ob Ostyaks; and another aid to this result is the adherence of the Ostyaks, at any rate nominally, to the Orthodox Church. They pay *yassak,* but are free from military service.

Social institutions.—A tribal system has never taken root among the Ostyaks. They are divided into clans each of which is really a large family. Groups of the same are ruled by a *knyaz* (prince) an hereditary office. The clans are artificially divided by the Russians into *radi* with territorial names.

Physical appearance and characteristics.—The Ostyaks as a whole are a people of only moderate stature, generally about 5 ft. 3 ins., to 5 ft. 4 ins. In type they are dolichocephalic, long-headed, as distinct from the Samoyedes who are brachycephalic, their heads being almost as wide as they are long. As the region of the Ostyaks approaches the Samoyede country, the brachycephalic type appears among them too.

The more northerly Ostyaks are not quite so well developed as those of the south. These latter are well built and have a distinctly thick-set appearance. All, however, look much the same; they have round, flat faces, broad and rather flat noses, projecting cheek-bones, dark, narrow eyes, and yellow or yellow-grey complexions. Their hair is long, smooth, and black, or chestnut coloured. There is very little hair on the face; in the northern regions the Ostyaks appear to pluck it out. The Mongolian type is more noticeable among the women than the men. The purest type is found among the fishers of the Ob, for the reindeer-keepers are largely intermingled with the Samoyedes.

In temperament they are kind, gentle, and friendly; they are very honest except in the neighbourhood of Russian settlements, where the contact of civilization has made them adopt ‘business methods’.

Knowledge and art.—In general the Ostyaks cannot read or write. In the Little Konda volost, some villages have been taught to read and write by discharged soldiers, and in the Atlimskoe villages on the Ob they are completely Russified. But the ordinary Ostyaks have now given up even sign-writing. They understand counting by tens, but have no knowledge of figures. They are skilled in handicraft, carving in wood and bone, and making beautiful decorations by scratching on bark, to ornament their vessels and baskets. The women make fine embroidery on linen woven by themselves from nettle, hemp, or flax. Hemp or flax is obtained from merchants; the nettle is gathered locally, dried, broken, and the fluff separated from the cover. Beautiful ornaments are also made with beads. The Irtish Ostyaks have a fairly good colour sense, through contact with Russians, but those of the north and east have little capacity for distinguishing between different shades. They reckon distances by the *êdip*, or Ostyak verst, which is five times longer than a Russian verst. For smaller measures, they have the fathom, which is the distance between the extremities of two outstretched arms, from finger-tip to finger-tip, or from the ground to the finger-tips of the upraised hand of a man of middling stature. They measure also by a span, the distance between the points of thumb and forefinger, and by the breadth of the middle of the hand, from forefinger to little finger. One hundred paces is sometimes expressed as 'as far as a marksman can shoot' (with an arrow). The smallest measure of time for the Ostyaks used to be the time required for a kettle to boil—something under one hour (see p. 165). Sunday is the only day of the week which has a proper name. The rest are numbered, 1st, 2nd, 3rd from Monday onwards.

The Ostyak's folk lore is tinged with sadness. They have an original music and poetry, improvised ballads accompanied with pantomimic action, and they also indulge in dancing. Their musical instruments are the *dombra*, which is a long instrument of pine-wood, shaped like a boat, with five strings of reindeer-sinews, and the 'swan', an instrument in the form of a bird with eight or ten strings of brass wire. Their idea of medicine is to bleed patients with an instrument consisting of a pike's jawbone fixed on a wooden shaft. They encourage even children to smoke tobacco, as a means of preventing throat and lung trouble.

Occupations.—In point of development, they stand between the nomadic and the settled stage of existence. They are not pure nomads, for most of them have fixed dwellings, which they inhabit in winter. On the other hand, they have practically no agriculture; only in the south in the *volosts* of the Narim and upper Demyanka do they till the soil. The rest of the Ostyaks maintain themselves chiefly by fishing and hunting, living in wooden or earth huts in winter, and leading a nomadic life in tents during the summer. A small proportion, in company with the Samoyedes, own reindeer-herds in the neighbourhood of Berezov and Surgut. The farther north they dwell, the more nomadic and less developed are the Ostyaks. They use tents in summer rather than huts and depend mainly on the reindeer for providing them with food, clothing, and the means of transport.

Although the lower Irtish lands are suitable for agriculture, the Ostyaks there, as elsewhere, devote themselves to fishing, hunting, and the collecting of berries and nuts. Such agriculture as there is, is done by Russian inhabitants. The harvest and fishing seasons synchronize; Russian peasant families engage in both occupations, by dividing the work among the various members of their family. But the Ostyaks are said to have too small families to do this.

Fishing, although very good on the Irtish and Ob, is not very highly developed among the Ostyaks. They use chiefly Russian fishing tackle, and as this is expensive, the villages or families to whom the best water or sandbanks belong, lease out these to Russians, for rents varying from 20 to 300 roubles (£2 2*s.* 6*d.* to £31 7*s.* 6*d.*) for a good fishing station; as much as 1,270 roubles (£135) has been paid for a sandbank in the Ob, with a neighbouring by-stream. Those who do not possess suitable sandbanks fish for themselves and gain 100 to 140 roubles per man. They catch sturgeon through the ice in winter by ingenious methods.

Hunting of the elk and reindeer goes on in the regions of the Irtish and Ob, chiefly in winter. The Ostyaks hunt on snow-shoes, with dogs, and use muskets of an antique pattern, often flint-locks, produced near Tobolsk at a price of from 2½ to 4½ roubles (5*s.* 4*d.* to 9*s.* 7*d.*). In remoter districts, even in the Irtish river system, bows and arrows were used at any rate in recent years, chiefly for shooting squirrels and ducks. Hunting brings in from 40 to 80 roubles per man. Forest-fires have greatly decreased its value. The skin of an elk brings in 5 to 6 roubles (10*s.* 6*d.* to 12*s.* 9*d.*) locally, that of a reindeer 1 to 1·50 roubles (2*s.* 1*d.* to 3*s.* 3*d.*).

The gathering of cranberries and cedar-nuts affords considerable employment to the Ostyaks, and it requires little or no capital. The woods are leased out by the village communities which own them, as are the fishing-stations.

The cranberry hedges are burnt out about every ten years, to get rid of old growths. The cedar-nut woods have suffered not merely from forest-fires, but from cutting for building-purposes. The best cranberry hedges are in the Konda district, where one man can gather 36 to 48 lb. a day, and one family during the season can collect 13 cwt. to 2½ tons.

In the eastern districts where the cedar-woods have suffered from fire and cutting, the economic condition of the Ostyaks has deteriorated. Formerly a family made £10 to £20 from collecting cedar-nuts; now in a good year, which means about one in four, they make £4 to £7. The average price is 2½ roubles (5*s*. 4*d*.) per pud (36 lb.). The woods belong to the village communities; any one can procure the right to gather nuts in a wood during the season for 1 to 1½ roubles (2*s*. 1*d*. to 3*s*. 3*d*.).

Horses and horned cattle are kept only for household purposes. Milk is used in the family; in the Irtish district a little is made into butter and sold.

Most of the good agricultural land in the Irtish district has passed into the hands of Russian peasants. The rest of the land is held by the village communities, which lease it out, but cannot legally alienate it, as it is Crown land and pays tribute. But as Ostyaks die out, the Russian Government acquiesces in their land passing to energetic Russian peasants.

Economic conditions.—There is a remarkable system of credit among the Ostyaks. Fish, meat, and berries do not supply all they require; they need in addition bread; tobacco, tea, and brandy are also in common use. A male Ostyak has also to pay something over six roubles (12*s*. 9*d*.) in direct taxes. To obtain ready money for all this, they could sell their wares in town, but the time and expense involved in transporting goods over a distance of, say, 60 to 200 miles is prohibitive. To meet this difficulty there are Russian traders who supply the Ostyaks with flour, tobacco, &c., also with money to pay taxes, sufficient for the year's needs. In return they receive the raw products of the Ostyak. At the end of the year, the value of these is reckoned against the value of the goods advanced by the trader. If there is a balance in favour of the Ostyak, it is given to him in goods ; if there is an adverse balance, it is carried over against him to the next year as a debt in the trader's books. The cheapest wares, such as fish, are only taken by the trader for sale on commission, in the town. The sum received is handed over to the Ostyak, minus the commission, and minus the cost of transportation, which is 10 to 15 kopeks for 100 versts, about 2½*d.* to 4*d.* for 66 miles. The Ostyaks cannot read or write, and therefore have to trust to the trader to keep his books properly. Among the northern Ostyaks, where the same credit system prevails, a notched stick is used for registering values ; the stick is split, and one half is kept by the creditor, the other half by the debtor. The Russian trader makes considerable gains, with a capital varying from 20 to 100,000 roubles (£2 to £10,500). But he may also suffer considerable losses if an Ostyak on his books dies in debt.

Ostyaks appear to be seldom free from debt. They quickly spend the money which they obtain for leases of river-stations, and then often become mere labourers at the rate of about 3 roubles (6*s.* 5*d.*) per month to those to whom they have granted leases.

Dwellings.—An Ostyak village numbers from four to twenty houses, arranged without any order. Besides such winter villages, there are summer villages, consisting of huts of beams roughly put together, used only during the fishing season. But the southern Ostyaks have given up the use of these, and make a living by residing continuously in their winter villages. Fishers who go off on fishing expeditions together, erect a temporary hut and live in it as if one family.

The fixed dwellings of the Ostyaks are of two kinds. To the north and east they are made of earth, but about Tobolsk and Berezov of timber. These last are of one, very occasionally of two stories, like those of the neighbouring Russian peasants. The windows are generally of glass , but sometimes only of skin. The houses have one, two, or three rooms, but in winter the whole family lives in one room, on account of the cold. Round this room are benches fixed to the wall; there are holy pictures, an enormous table, a great wooden bedstead, and a stove built of clay. The earth-huts of the other Ostyaks are of the same pattern, but on a much smaller scale. Near the winter dwellings of the Ostyaks are sheds for stores. The horses are kept away from the house, generally in the open, sometimes in a wooden shelter; horned cattle are provided with a shelter.

Clothing and food.—The Irtish Ostyaks are in the habit of wearing Russian apparel, except that in winter some of them adopt the Samoyede costume, as indeed Russians do too. This consists of an outer garment and an inner garment (*malitsa*) both made of dog-skin or reindeer-skin, and a fur cap. The rest of the Ostyaks wear much the same dress as this in winter. In summer they wear only one skin-garment with another of red cloth above it. Leathern trousers, leathern stockings, a belt with requisites hanging from it, and shoes of reindeer-skin complete their costume. The women wear stockings made of fish-skin: otherwise their dress does not differ from that of men, except that they have a piece of drapery with which to cover their faces.

For food they eat mainly reindeer-flesh and fish, preferring to eat them raw. They regard the former as a great preventive of scurvy. *Poziom* (dried fish) and *varka* (the stomach and entrails of fish soaked in oil) are much eaten by them. From Russian merchants they have learnt the use of bread, and it is spreading among them. They make what is called *burduk* of meal boiled with water and fish-bones.

Marriage and death.—Marriage is exogamous, though the clans are not composed of blood-relations. A price is paid for the bride.

The Ostyaks are buried in forests. No grave is dug, but the body is laid on the ground and covered with inverted skins. In the north they still bury the dead in canoes.

Religion.—For about a century the Ostyaks have been baptized, and Christianity has made some progress among them. On the Irtish and Ob the old customs are largely forgotten, but in the north the nomadic reindeer-breeding Ostyaks are still pagan. They recognize good and bad gods, and pay homage to the god of the thunder and the spirit of the River Ob.

A special sanctity attaches to the bear. They swear their most solemn oaths over its pate, and when they have killed and eaten one they are careful to collect and bury its bones. The loss of one will entail some mischance at the next bear hunt. They apologize to the bear for killing him, saying that the real responsibility rests not with them, but with the Russian who supplied the gun and gunpowder.

(ii) Samoyedic Tribes

Samoyedes

Territory.—This important tribe extends along the north coast of European and Asiatic Russia from the shores of the White Sea to the banks of the River Khatanga. They are most numerous in the province of Arkhangel. In old accounts of voyages Samoedia was always the country between the Pechora and the Ob, but they really extend almost to the Lena. They have come northwards from the Altai, driven out by the Turco-Tartars in the fifth century A. D., and they may be the people traces of whose primitive civilization are found in the valley of the upper Yenisei. Their present mode of life recalls the ‘ reindeer-civilization ’ of primaeval times in middle Europe.

Name.—There are various spellings of the name : the form adopted in this book is that most frequent in literature, but the pronunciation seems to be more like Sam-yad. The word has been supposed to mean ‘ self-eaters ’, or has been otherwise explained as ‘raw-eaters’. Probably, however, it has no such barbarous connotation, but is to be associated with *Suomi,* the name by which the inhabitants of Finland call their country. The Lapps and Karelians also have similar names for themselves. The Samoyedes call themselves *Hazovo* (i. e. the men), and *Nyänyäz* ; the Ostyaks call them

Orghoy and *Vorkho*, names which recall the word Ugrians, by which name the inhabitants on both sides of the Ural are called. The name Samoyede occurs in a chronicle of 1096.

Racial affinities and language.—They are usually distinguished from the Finno-Ugrian tribes, but are closely allied to them. The language resembles Finnish, but has more suffixes : it is agglutinative and polysyllabic, and is sonorous and pleasant to hear. There are three dialects and twelve sub-dialects.

Numbers.—It is difficult to be at all sure of the numbers, as the methods and dates of computation in Europe and Asia differ. A rough estimate of the numbers is about 20,000. In Arkhangel province in 1897 there were computed to be 6,748, and in Asia in 1912 they were estimated at 12,502.

In Novaya Zemlya there are about 100 Samoyede settlers.

Another reason for caution in dealing with the question of Samoyede numbers is the uncertainty whether they include the southern tribes closely allied to the Samoyedes. There has been a slight increase at any rate in Europe since the middle of last century, and probably the gloomy prognostications that the race will die out are unwarranted, as they are better adapted to the land in which they live than any one else. But their marriages are not very fertile, one mother seldom having more than two or three children, and they have to face four deadly foes, syphilis, scurvy, small-pox, and spirits.

Divisions.—Of the Samoyedes proper there are four main divisions :

(1) The *Yuraks*, who extend from the White Sea to the Yenisei, a vigorous, brave people, who are mostly reindeer-nomads, but who also do some hunting and fishing and show great daring in their expeditions. Their chief centres are in the neighbourhood of Obdorsk, where there are about 6,000 of them, and near the Gulf of Mezen, where there are about 5,000. There are five sub-dialects of their speech.

(2) The *Tavgi* live in the Taimir Peninsula from the Yenisei to Khatanga Bay. There are about 1,000, most of whom are reindeer-nomads.

(3) The *Ostyak-Samoyedes,* whose number is uncertain, but who are estimated at about 3,000, live in the zone between the tundra and the taiga. They are almost entirely hunters, there being merely a few reindeer-nomads in the northern part of the district, and they own but few reindeer.

(4) The *Yenisei-Samoyedes* live along the Yenisei : they live mostly by fishing, do some hunting and only to a slight extent are reindeer-nomads. There are only about 350 of them.

There is further subdivision into tribes : for instance in the Yamal Peninsula there are ten different tribes, each with its fixed boundaries for reindeer-pasture.

Relations with Russians.—Their connexion with the Russians, at any rate in Europe, is of long standing : in the eleventh century we hear of them paying tribute to the Novgorodians. Their present relations with their Russian masters are believed to be good, but a readiness to do work without pay suggests, apart from their natural willingness to oblige, traditions of forced labour. They have been protected in their occupation of the tundras by law since the sixteenth century, and they resist and resent encroachments on the part of the Russians and Zirians, but, being bad men of business, they easily fall into a position of dependence upon them. They have been impoverished in the south by the loss of their hunting-grounds, as Russian civilization has spread northward. On the Yenisei the fisher-folk are less independent than the other Samoyedes, and the Russian traders are apt to be despotic with them. But there is certainly no general desire to grind them down or treat their rights disrespectfully : the adult Yurak pays an annual tax of about £1, which is not a very heavy impost.

Social institutions.—In 1835 the Samoyedes were given considerable powers of self-government. In their own affairs they are governed by *starshinas* (elders or mayors)—the name has generally replaced the earlier one of *knyaz* (prince)—one of whom is elected for each tundra. He is the intermediary between the Samoyedes and the Russian administration : he is the ruler and judge of his little community, collects the *yassak* (tribute) and pays it in to the Russian Government. All offences except the most serious are settled according to their own customs. A further law of 1892 recognized and sanctioned native councils; the *munyak* meets annually, their meetings being held in winter; there must be a representative of each clan present; women are excluded from the meetings. The *starshina* is elected for three years. Russian influence has greatly strengthened his hands, and his power is probably greater than when he had the more exalted title.

Physical appearance and characteristics.—Accounts of the temperament and stature of the Samoyedes differ as much as do those of their numbers. It is probably difficult to generalize about a people so widely extended and so markedly divided. In height there is a general agreement that they are short: a man 5 ft. 6 ins. would be accounted a giant, and the men are about 4 inches taller than the women. In appearance they resemble the Ostyaks, but they are brachycephalic, or rather, mesocephalic. They have straight, glossy, black hair, which is usually bound with thongs in two bunches; their skin is sallow; their eyes narrow, oblique, and far apart; their faces broad, flat, and round; the nose flat and open; the cheekbones prominent; the lips thick. They are mostly beardless. They are often of good appearance despite their short stature, are stoutly built and very muscular.

There is some admixture of Russian blood, and the race is found to be less pure in the west than in the east. On the west side of the Pechora a Slavonic and Teutonic strain shows itself in the breed: so there are some Samoyedes in these parts with light hair, fair skins, and eyes 'of Gothic type'.

Travellers differ as to their honesty, but the general verdict is in their favour: at any rate they are known to adhere strictly to their word when given. They have energy and natural intelligence, but are thriftless. They are sociable, and extremely hospitable, both to strangers and their fellow tribesmen: they are constantly smiling and laughing, and delight in gossip; they are very fond of children, who treat their elders with confidence and without fear. They are more independent than the Ostyaks, and the Yuraks especially are brave and daring. They are a dirty people, and never wash ; nor do they change their clothes, until they are worn out. But their teeth are white, partly because they are accustomed to chew pine-resin. They are fond of music, though their music is much less developed than that of the Ostyaks, being rudimentary and monotonous. But the Samoyede likes to possess himself of the *dumbra* of the Ostyak.

Occupations.—The chief occupations are reindeer-breeding, hunting, and fishing ; before their northward migration they practised agriculture, but very few practise it now. The Yuraks and Tavgi are mainly reindeer-breeders, and, as such, nomadic ; but the Samoyedes are capable of settling down. At Kozhva, in the Arkhangel province, where there is some stationary population (in 1892 there were 38 huts) the people are engaged partly in growing barley, partly in rearing cattle, partly in fishing. But the Samoyedes who have 'settled down' often live in Russian and Zirian villages without a regular occupation. The nomadic Samoyede is on the march with his tent for four months every year ; he migrates south in winter and north in summer. A rich Samoyede—the richest are in the Yamal Peninsula—may have as many as 5,000 reindeer, and the further north one goes into the tundra, the richer are the Samoyedes found to be. They seem to have no affection for their deer, but regard them merely from the pecuniary point of view. Some of them, especially the Ostyak Samoyedes, are more engaged in hunting, and to some extent in fishing. Communal hunting of the wild reindeer still survives among the Samoyede. White and blue foxes are trapped and snared, and geese and other birds are caught during the moulting season. In hunting they used to employ a bow and arrows, but in their place they now use clumsy and primitive flint-lock muskets, and employ a gun-rest in shooting. So a native industry in bows and arrows has died out without any counterbalancing addition to their resources. The fisherman is looked down upon by the reindeer owner. There is a weak industry on the sea coast in hunting marine animals; but the incapacity and lack of enterprise of the Samoyedes put them at the mercy of monopolists: they do not even carry their goods to market, but have them transported by the Russians, who use the Samoyedes' own reindeer for the purpose. Most of the Samoyedes' implements are of bone and stone. But with three metal tools, his axe, his borer, and his knife, he is very dexterous. The Samoyede women are expert in sewing: they use reindeer-sinews for thread, preparing them first by chewing.

The Samoyedes trade in much the same way as the Ostyaks (p. 143): they bring their peltries to Berezov, Obdorsk, and other markets; but the Zirians, whom the Russians call the Jews of the tundra, have got most of the Samoyede trade into their hands. East of the Urals money is little used, but in bartering with the Samoyedes only articles of practical use are serviceable: to think of them as savages and bring them glass beads is the worst of mistakes.

Dwellings.—The migratory Samoyede lives in a *chum.* Three or four families usually travel together. A *chum* is made of about twenty fir-poles, sharpened at each end, driven into the ground and with their tops lashed together. Over these are tied large pieces of birch-bark, reaching from the top of the poles to the ground and secured in their place by stones or lumps of earth. The chimney is an orifice of two feet or so between the tops of the poles and the edge of the birch-bark. In winter the *chum* is covered with reindeer-skins, well caulked with moss. In the centre of this tent is a large flat stone on which the fire is made. The fire serves for illuminant as well as heat, for they do not use lamps. The part of the *chum* that faces the entrance is holy, and must not be contaminated by the presence of women, as they are unclean.

Clothing and food.—The common dress of both sexes is a red cotton shirt and thin cloth trousers, with skin stockings (*luipti*) and long deer-skin boots (*pimi*) which are almost alike in the costume of men and women. In snow seal-skin boots are worn instead as being more waterproof. Over the shirt is worn the *malitsa*, a smock-like garment made of reindeer-skin, with the hair worn inside ; it has a hole for the head to be thrust through, and at the neck there is attached a close-fitting hood, while mittens (*rukavitsa*) are attached to the sleeves. It is tightly girded round the waist, so as to make a sort of bag. Over this is worn a white deer-skin (*suvik*), cut in the same pattern, but without the *rukavitsa* ; instead of these it has bands of red flannel about the wrists : in this garment the hair is worn outside. Women do not wear the *suvik* but 'a long, loose, buttonless skin-coat, reaching to the calf of the leg, folded over the breast and secured round the waist by a belt. At regular intervals there are eight or nine strips of reindeer skin with intervening pipings of red and green flannel'. The woman wears a cap detached from the robe with hair outside and elaborately adorned. The only ornament in male attire, as a rule, is a belt of thongs with metal buttons : both sexes wear charms, especially the tooth of a bear.

Their principal food is reindeer meat, which they like to eat raw or half-putrid. They care more for quantity than quality. One favourite delicacy with them is the gullet of the reindeer. The young reindeer is good to eat, but the old reindeer, if not hung, is very tough. They are fond of *vodka*, and are ready to sell a reindeer for three litres of it. They chew tobacco, but do not smoke it, but they make snuff by grinding down tobacco. In some parts a certain amount of rye-bread is made, where the flour is mixed with water and fermented.

Marriage and death.—Monogamy is general, but there is no objection to polygamy. Few Samoyedes, however, have more than two wives, though some rich men have as many as four in separate *chums*. The father only keeps part of the *kalym* paid by the bridegroom, which may amount to as much as thirty fox-skins and three hundred reindeer (an actual case) ; the rest is given to his relations. He gives as a dowry a *chum*, some reindeer, sledges, harness, meat, and clothes, amounting in value to the kalym paid, which is returned in case of divorce. A Samoyede will sometimes sell his wife for some teams of reindeer or exchange her for the wife of another man.

Interment of the dead has been the custom with the Samoyedes since their conversion to Christianity. Graves are viewed with superstition, and the head is averted in passing the grave of any one who is not a relative. The graves are merely rude wooden boxes, often rifled by wolves and foxes. Beyond the mouth of the Ob below the earth lies the world of the future life, where the shadow will live as long as it has lived on earth while the soul is reincarnated.

Religion.—A 'conversion' to Christianity, which had been preceded by a considerable assimilation of Christian ideas and ethics, has not prevented the retention of a good deal of the primitive religion of the Samoyedes. They have a feeling that the God of the Christians cannot be supposed to know much about reindeer; so in that part of their life which is related to their herds, they make their appeals to their native divinities. At the head of these is *Num*, the giver of life, a highly exalted being who will not even deign to glance at the earth, as being unclean. Of him there are images made called *chaddi*, which are carefully kept out of sight. Besides Num, they believe in *Aa*, a devil, *tadebtsi*, spirits, and *hegi*, household gods. These last can be approached directly without a mediator, but for the *tadebtsi* the intervention of the *tadebeys* (Shamans) is necessary; they are the embodiment of the divine spirit on earth. There is no particular temple for their rites, but they resort to certain places on hills. The island of Vaigach is accounted especially holy, and there above all places the devout Samoyede would wish to be buried.

Tribes akin to the Samoyedes

Besides the Samoyedes, properly so-called, there are tribes further south, closely akin to them. Of these may be named the Beltirs, Kaibals, Kamassins, Karagasses, Motors, and Soyots. Some of these are largely mixed with a Tartar strain. The Beltirs live by agriculture and cattle-breeding on the Abakansk steppe; they profess Christianity; their language is like that of the Sagai Tartars. The Kaibals are on the upper Yenisei; they are hardly to be distinguished from the Tartars of the Minusinsk district; they support themselves by rearing cattle. The Kamassins, who also have a large Tartar admixture in their language, live in the Kansk district of Yeniseisk. They are herdsmen or agriculturists. The Karagasses, north of the Sayansk Mountains, are losing their distinctive features. They number to-day 345, including 83 'warriors'. Of the Motors one section entered China and was exterminated; the other section has been merged among the Tuba and Soyots.

Soyots

The Soyots are probably a division of the Samoyedes, who live in the extreme south of Siberia near the Kitoisk Mountains. They are a small and rapidly disappearing tribe, isolated from the rest of the world among their cold and bare mountains. Their settlements are at Lake Ulchir, Kitoi, Tsagan-Khar, Oplik-Gol, Samart, Dzhatkhak, and Khonshon. They live entirely by hunting for sable, squirrel, and bear, and procure their own requirements (such as powder, lead, tea, flour, and salt) by the sale of their fur. They make use of horses, but more often of reindeer, for driving and hunting. They live in wooden *yurtas* in different places in accordance with the season of the year, the winter ones being built on the edge of the taiga so that they can procure firewood. The position of the others is dictated by the necessity of procuring food for their horses and reindeer, which have to graze all the year round, as they have no means of making hay near their *yurtas*. These huts are four-cornered erections of wood, with an opening in the roof for the smoke from the fire which is always burning. They live amid incredible dirt and never wash, their only method of cleaning being by licking. Their religion is sometimes Lamaism, sometimes Shamanism. They speak the Buryat language. A few years ago there was one old man who could speak the Soyot language. Some few of them speak Russian. Their one food is *zamuran*: i.e. green brick tea, boiled with milk, flour, reindeer-fat, and salt; they never eat bread and only rarely reindeer meat. Their implements for eating are a small wooden cup of Mongol workmanship, which the owner always carries about with him, and a knife, without a sheath, hanging from the belt. They suffer much from various diseases—colds, complaints of the stomach, and syphilis.

(iii) Turkic Tribes

The Turkic tribes have often been classed with the Mongolian, but it is now more usual to separate them. The differences are more marked in language and religion than in physiognomy and other racial characteristics. There is a considerable blending with Finnic and Samoyedic stocks, and in some cases it is difficult to say to which of these branches of the Ural-Altaic group a particular tribe like the Karagasses should be attributed. In western Siberia the great group of Turkic peoples is the Tartars, while in eastern Siberia their chief representative is the Yakut. The Turkic people have penetrated Siberia from northern Mongolia, preceding in their movement the great Mongol invasions of the thirteenth century. It was these invasions which caused the Yakuts to move from the region in which they had settled round Lake Baikal to the basin of the Lena. The Tartars spread farther west, and there are well over a million of them in European Russia alone.

Siberian Tartars

Territory.—In Siberia the Tartars are found west of the Yenisei. They are cut into two distinct portions; a northern group, consisting of the Baraba and Chulim Tartars, occupies the Baraba steppe and the Chulim basin to the east of it, and a southern group, separated from them by the Siberian Railway and the stream of Russian immigration, occupies the Altai and the Abakansk steppe in the basin of the Yenisei.

Name.—There are various names for the different groups of inhabitants, but they are collectively called Tartars, or, with more correct spelling, Tatars.

Racial affinities and language.—As stated above, they are distinct from the Mongolians, though sometimes classified with them, and many of the tribes have become much intermixed either with Finnic or Samoyedic peoples or with other peoples of Turkic stock. In some cases too they have grown to have a strong resemblance to the Caucasian peoples, with whom they have intermingled.

In language those of them who have kept their own speech resemble the Ottoman Turks, but some tribes, like the Chulim Tartars, have lost their own tongue, and others like the 'Kalmuks' of the Altai have a large number of Mongol words in their vocabulary.

Numbers.—The number of Tartars in Siberia is reckoned (1912) at 176,124. They constitute nearly a third of the Turkic stock in Siberia.

Divisions.—North of the Siberian Railway, in the district between Tobolsk and Tomsk, are two Tartar peoples: (1) The Baraba Tartars, of whom there are about 55,000, live in the Baraba steppe between the Irtish and the Ob. They are an agricultural people, who have given way before Russian encroachment, and live now in villages among the marshes and woods. (2) The Chulim Tartars are very few. They live on the River Chulim, a tributary of the Ob: they have almost entirely given up the use of their native speech, and live for the most part like Russian peasants.

South of the Siberian Railway there are two main groups:

(1) The Altaians, of whom there are various subdivisions: the Tartar Kalmuks (about 12,000), who are not really Kalmuks at all, as the real Kalmuks are Mongolian, which this Altaian people are not despite a Mongoloid appearance; the Teleuts or Telengites (about 5,800) in the Kuznetsk district; the Chern or Black Forest Tartars, farther north on the River Biya; the Shors, about 11,000, on the Rivers Tom and Mras-su; and the Lebed Tartars, along the River Lebed.

(2) The Abakansk Tartars, on the Abakansk steppe in the valley of the upper Yenisei, in the neighbourhood of Minusinsk,

have become much Russianized and lost many of their national characteristics.

Relations with Russians.—Many of the tribes are adopting Russian costume and habits, and some like the Chulim Tartars are losing their native speech. On the whole relations between the Russians and Kalmuks are friendly, though extensions of the activities of Russian tax-gatherers are sometimes actively resented. The Kalmuks, probably as so closely resembling the Mongols in appearance, are not—or at any rate were not lately—allowed to perform military service for the Russians, though it is likely that they would make good soldiers.

Social institutions.—The Altaians are divided into clans, but the separation is not very marked, and people of various clans live together in the same village. In this they differ very much from their western neighbours, the Kirghiz. The clans are known as *seoks* (generations), of which there are supposed to be twenty-four, but it is doubtful whether the number has more than a mystical significance. People of the same *seok* regard themselves as related to one another.

Physical appearance and characteristics.—The Altaians, if they may be taken as representative, have flat, broad faces, small foreheads sloping backwards, little eyes turning upwards at the corners, small eyebrows, prominent cheek-bones, snub-noses too small for the face, large mouths with thick lips, which display two rows of strong white teeth. The chin is pointed, and they have little or no beard. Their complexion is dark, their hair and eyebrows deep black, the hair being stiff and bristly. As a rule the men shave off a good part of it, leaving only a patch on the crown of the head, which they plait into a queue. They tend to be short and broad-shouldered, but do not look very strong, probably because of the poor nourishment of a large number of the people. Their legs are bowed—at any rate this is the case with the Kalmuks—because they have lived in the saddle from their earliest years. The women are smaller than the men, but look stronger: it is they who do a great deal of the hardest work. They have a general reputation for being honest and industrious. They are also an imaginative people, and the Kalmuks have a considerable store of legends; they are exceedingly fond of their own regions, as their songs show. Their chief vices are dirtiness and drink.

Occupations.—They are mostly nomadic, and pursue a pastoral life, but the Baraba Tartars engage in agriculture. The Kalmuks are great equestrians, and are brought up to the art of riding from their earliest years. They carry most of their possessions about in sacks on their horses' backs. There is a peculiar breed of horses in the Altai, but it is seldom found pure now, being much mixed with Russian, Kirghiz, and Mongolian stocks. The Altaian horse has a beautiful, erect, 'dry' head, large eyes, a deep curb-dimple, a broad throat and finely set nervous ears. The neck is dry, sinewy, and short in proportion to the length of the animal; the chest is broad. The withers are not particularly high, and die away with the back. The shoulders are slanting, the upper part of the thigh strong and sinewy, the knee broad, the pastern steep and short: the hoof is steep and small, very firm, and except for a large frog, has a normal form. Most have a bright colour. The height is 4 ft. 2¾ ins., the length 4 ft. 9 ins. Besides its beautiful bodily form, the Altaian horse is distinguished for speed and cleverness.

They also have a special race of horned cattle: the cows give plentiful good milk. The Altai Mountains are a paradise for cattle-breeding, the numerous streams and short grass offering good nourishment for horses and horned cattle; there is also good pasture for sheep and goats. There are no harmful insects, and the pastures are free from snow in winter, so that the animals can get their own food.

Dwellings.—The Tartars live in *yurts*, which are like those of the Kirghiz, but their villages are less exclusive than the *auls* of the Kirghiz, more families dwelling together.

Clothing and food.—They wear a sort of shirt with a cape; the sleeves of the shirt are long, reaching half-way down the hand; the shirt is held together by a girdle, and is made of some blue Russian cloth, or blue or brown Chinese material. They wear trousers of the same material, wide and reaching just below the knee. Sometimes the trousers are made of roe-leather. The shirt hangs free over the trousers, just below the waist. The footwear consists of shoes without heels, reaching to the knee, made in summer of dressed sheep-skin, and in winter of undressed skin with the hair outside. They wear felt stockings projecting about two inches above the stock of the shoe, and between the stocking and stock of the right leg they put their tobacco-pouch and pipe with the stem projecting. Over their shirt they wear a cloth jacket, with sleeves reaching to the elbows, and with long pockets hanging down. Above this they may wear a leather coat in summer or a skin coat in winter, with a belt holding a pouch, knife, &c. On their heads the Altaian Tartars wear a three-cornered hat, the point being behind, with erect rims. It is made of lamb-skin, and is covered with some yellow material, with a red oval flap sewn on the top. From the corner behind hang two red ribbons, one and a half feet long. They eat mutton, whether the meat of the domestic sheep, or of the big-horn that frequents the Altai; they are also fond of the flesh of horses. Though they own large herds of cattle, they do not eat them.

Religion.—The Baraba Tartars are Mohammedans. The Southern Tartars are either Christian or heathen. One of the outward and visible signs of Christianity is the absence of a queue on the top of the head; another is monogamy. They have many superstitions, and Shamanism is prevalent among them. Polytheism is common.

Other Tartar Branches

In the above description the Altaians and more especially their western branch, the Kalmuks, have been dealt with. A few words will suffice about the other branches. The Teleuts in the eastern Altai have much the same appearance as the Kalmuks, but they are more settled, and engage in agriculture. Their language and poetry is of much the same character.

The Shors live mostly by fishing, and are very poor. In language, religion, and costume they have become much assimilated to the Russians. Closely resembling them are the Lebed Tartars. The Chern Tartars live by hunting, trade, and cattle-breeding. One profitable product of their land is wild honey. They also sow barley and wheat. They are extremely poor.

Kirghiz

In addition to these tribes of the Altai and its northern foreland, there is the important Turkic stem of the Kirghiz, with its two great branches, the Kara-Kirghiz and the Kazak.

Kara-Kirghiz

The first-named, whose appellation means 'black', are really outside the area described in this book. They are found in great numbers in the government of Semiryechensk, especially in the neighbourhood of Lake Issuk-Kul and in the steppes south of Lake Balkhash. They number in all about 800,000, of whom 700,000 are in this government. They are governed by tribal rulers, elected by themselves, who enjoy unlimited authority and with whom the Russian administration interferes but little. They carry on a vigorous trade in live-stock breeding—horses, cattle, sheep, goats, pigs and camels.

Kazak-Kirghiz

Territory.—This great people are found from Lake Balkhash to the Aral and Caspian Seas and to the lower Volga, and also in the regions of the upper Irtish and Ob.

Name.—The name by which they are always known among themselves is *Kazak*, which means 'rider' and is the same word that we know in the form Cossack. But the Russians, since the word Cossack has come to mean something very different, call them loosely, but conveniently, Kirghiz.

Racial affinities and language.—They should be carefully differentiated from the Kara-Kirghiz, though they share so many of their customs and ways of life, for they differ in physiognomy and language. The Kara-Kirghiz seem to have come from the Altai, the Kazak-Kirghiz from Asia Minor. Their language is Turkic in structure, but contains a large number of Mongolian, Arabic, and Persian words.

Numbers.—They are said to number between two and three millions, and increase steadily as one goes further south.

Divisions.—They are divided into three Hordes, which are subdivided into races and tribes. The Great Horde lives in the region south-east of Lake Balkhash from Semipalatinsk to the Ala-tau Range, the Middle Horde occupies the watershed between the Aralo-Caspian basin and the River Ob, the Little Horde is mainly between the Aral and Caspian Sea. A fourth Horde, called the Inner Horde, has been settled since 1801 in the Orenburg district. The higher orders of the Kirghiz are divided into White Bones (Ak-sijuk) and Black Bones (Kara-sijuk), according as they are descended from khans and 'saints' or other ancestors.

Social institutions.—They choose their own *khans*, who have authority in their respective tribes, but little beyond them. These appointments are confirmed by the Russian Government. The real rulers are the elders, who are appointed by public election. Rigorous punishment is dealt out to the brigandage and raids which arise from inter-tribal feuds.

Physical appearance and characteristics.—They are allied ethnically to the Mongolians, and preserve strongly marked Mongolian features though with admixture of Finnish and Iranian blood. They are middle-sized, square-built, and inclined to be stout. They are brachycephalous with small slanting eyes, which are usually black, high cheekbones broad, flat nose, small mouth, long black hair, and very little beard on the face, which is usually of a yellowish-brown hue, but occasionally fair. Their hands and feet are small. They are an honest and trustworthy people, but have a sullen and unfriendly manner, and more so, the nearer that they dwell to civilization. They are more warlike than the Kara-Kirghiz. They are a hardy, long-lived people, but suffer from dirt, small-pox, ophthalmia and syphilis. They are very fond of recitation, but have no dances. Their musical instruments are the *kobys*, a string instrument, and *suvusya* (a kind of shawms).

Occupations.—The Kazaks have vast flocks and herds, which constitute their main wealth, and they are admirable horsemen. They are employed in the mines and in fishing.

Dwellings.—They are nomadic and have no settled homes, though apt to fix their summer tents year after year at the same spot, when they are driven by drought and insects to the upper mountain pastures. They live in *yurts*, which are circular tents of light wooden framework with coverings of red cloth or felt, and a hole at the top for light and ventilation. The whole can be set up or pulled to pieces in a few minutes : its weight is about 10–12 puds (360–430 pounds), and it makes a load for two camels. The height is from 8 to 15 ft., the diameter from 10 to 30 ft. The winter *yurts* (*zimovkas*) have coverings of felt two or three times as great. An invariable feature of the *yurt* is the *kazan* or large iron pot in the centre.

Clothing and food.—The ordinary dress is a *chapan*, a flowing robe ; the number worn is determined by the season of the year and the material by the wealth of the wearer. This is fastened by a girdle of silk or leather, which contains his knife, tobacco-pouch, and other inseparable accessories. Broad silk pantaloons are worn, and black or red leather boots. There is little difference in the costumes of the two sexes. A *malachai* envelopes head and neck and most of the face. Sometimes three pairs of shoes are worn : first a loose pair of thin embroidered leather boots (with moccasin soles), then felt boots (*pimi*) of solid wool about one inch thick, and then stout leather boots, all coming well above the knee. Sometimes there is also worn a stout overcoat reaching to the knee, lined with curly-wool sheep-skin.

For food they devour horseflesh and vast quantities of boiled mutton. Instead of bread they eat *balamyk*, a mass of flour fried in dripping and diluted in water. They drink *kumis* (fermented mare's milk) in large quantities.

Marriage and position of women.—The women are restricted less than is usual among Mohammedans. It is customary for a Kirghiz boy first to marry a woman older than himself, to guide his footsteps in married life ; some ten years later he weds a younger wife, and his earlier spouse is relegated to the position of dowager.

Religion.—Like the Kara-Kirghiz, the Kazak-Kirghiz are Sunnite Mohammedans, but their belief is of a lax kind and largely tinged by Shamanism. They have few mullahs, and make few pilgrimages to Mecca, though they revere the tombs of their local ' saints ', which are a prominent feature in the steppe landscape. They have a month of fasting called *Urazah.* The chief prescription of the Koran for which they show respect is polygamy, but that is restricted by their limited means.

Yakuts

Territory.—The Yakuts occupy the valley of the Lena for almost its entire length : in the north they extend from the Khatanga to the Kolima, in the south from the sources of the Vilyui to the Sea of Okhotsk.

Name of tribe.—They call themselves Sokha or Sakhor : there is still in the neighbourhood of Minusinsk a Tartar tribe called Sekho, who may have retained the name from a time anterior to the northward migration of the Yakuts. They owe their name Yakut to their Tungus neighbours who call them Yako or Yakot. One branch is found between the Yenisei and Khatanga Rivers in the Government of Yeniseisk, who are known as Dolgans or Tolgans : they are a fishing, nomadic folk.

Racial affinities and language.—The Yakuts belong to the Turkic branch of the Ural-Altaic stock, and their language presents close resemblances to that of the Turks. There is a Yakut grammar by Boethlingk (published in 1851).

Numbers.—Including the Dolgans, they are computed at 226,739, of whom 113,330 are males. Their marriages are fertile, with an average of about ten to a family, but less productive as one goes north. They are increasing in numbers and absorb a large number of Russian settlers, who learn their language and adopt their customs.

History and relations with Russians.—They were expelled from the Baikal region about the thirteenth century by the Mongol Buryats, and driven northwards. They had to go far north to escape the depredations of their Tungus neighbours, but they have now proved themselves considerably their superiors, and their settlements are extending south again, while the Tungus are giving way before their superior civilization. The Cossacks found them in 1620, at a time when they were the prey of internal dissensions.

Social institutions.—They are divided into *uluses, naslegs,* and *aga-usa* (clans). Over the *ulus*, the biggest division, presides the *golovà* with an *uprava,* a kind of police-court; several *naslegs* go to make up the *ulus*: they are administered by a district council under a district elder called *kujas.* The *nasleg* consists of from one to thirty clans, the clan sometimes being composed of a few individuals, sometimes of several hundreds. The council of elders of the clan, which used to decide all legal and economic questions, was called *ogonyor.* The Russians used to tax the Yakuts in furs; now they tax them in coin. A poll-tax of 4 roubles is or was levied on them, the richer among them paying income tax as well. The clans were originally very large, when the Yakuts owned great herds of horses; now that cattle-rearing predominates, the herds and the clans have become smaller. A clan by its own laws is responsible for the crimes committed by its members; sometimes one clan makes an alliance with another, celebrated by joint sacrifice and festival (*ysyakh*). Marriage-gifts are also a symbol of reconciliation between clans, and compensation for damage done in their feuds.

Physical appearance and characteristics.—The Yakuts differ from one another in physique more than any other Turkic or Mongol people. Those that are well-off and can get enough to eat are from 5 ft. 10 ins. to 6 ft. tall: they are well-proportioned with good chest-measurement, and are robust and active; but in the north, owing to severity of climate and badness of food, we find a very different type: they are below medium height, with a sickly complexion, and, unlike most Yakuts, are indolent. Ordinarily they are thick-set, robust, and muscular. They have black wiry hair, which the men cut close to the head, except the Shamans who grow it long and tie it up behind, dark and elongated eyes, broad flat noses, narrow foreheads, small round heads (the men's faces are long, the women's oval), little beard. The women are somewhat ungainly, and do not add to their attractions by the paint and excessive ornament that they affect, and the adoption of European costume still further diminishes their charms. The mode of speech of the Yakuts is slow, disconnected and abrupt. They are described as good-tempered, orderly and hospitable, with more energy than their countenances suggest. They are capable of enduring a long strain, and are patient under privation; but they lack independence, and need stimulus: it is said of them: 'the more you beat them, the better they work.' They are apt to fall into debt, mostly to their own fellow-countrymen, for the more enterprising among them have earned the reputation of being the Jews of Siberia. They are fond of noise, song and dance; their improvised songs have deadly monotony as they consist of eternal repetitions of one phrase.

They measure distance in terms of time; for instance a *kess* is the distance done in the time it takes a kettle to boil, a sufficiently elastic term (see p. 141).

They are passionately fond of cards; when giving their horses rest on their journeys, they instantly produce their cards, and are ready to stake anything. They are great eaters and drinkers, great smokers of tobacco, and extraordinarily fond of sleep. They are probably the most intelligent of the Siberian native tribes, though the Buryats are better educated. They carve figures of human beings which they fashion out of mammoth-tusks: but these are rigid and lifeless, with only an outward resemblance to what they represent, and much inferior to those of the Koryaks. Even these specimens of their plastic art are rare and merely made for commercial purposes, to sell to the Russians. They take no pleasure in them themselves.

Occupations.—Their principal occupations are looking after horses and cattle: the former has been their industry for a very long time, the latter has been more recently adopted. They also hunt and fish, and search for fossil-ivory. Cereals have been introduced into their country and are becoming acclimatized. They smelted the iron-ore of the Vilyui valley long before the Russians came, and they manufacture their own tools. Yakut steel is very flexible, but very good. They have only lately begun to breed dogs. They regard the dog as an unclean animal, and as having no *kut* (soul); they are harsh to their dogs: it is a gross offence against good manners to introduce a dog into a Yakut's house 'because of the evil spirits that sit on its tail'. Their habit of horse-breeding led to them setting the *yurtas* at a considerable distance from one another, because it was necessary to have plenty of pasturage for their big herds; if they had exhausted the pasturage near home, it would have been necessary to send the horses a long way, which would have meant risk from the negligence or dishonesty of the keepers.

Dwellings and furniture.—Their winter-dwellings are *yurtas* made of logs: the doors are made of raw hides, the walls of logs or wicker, caulked with cow-dung and flanked with banks of earth to the height of the windows. In place of glass they have in the windows either skins or sheets of ice which are kept in place by slanting poles, and frozen afresh into their positions at intervals by pouring water round the edges. The roofs are flat and covered with earth. The door faces east, and is protected by projecting boards. Inside the *yurta* is divided into two 'apartments': one is for the cattle, the other divided into sections for the family. In the middle is a round chimney made of wood, but guarded against risk of fire by a thick covering of clay. In this chimney are placed cross-beams, from which they hang their kettles and pots. Their chief cooking utensil is a large iron pot; this was so valuable when the Russians invaded their territory that the price demanded for it was as many sable-skins as would fill it. In the summer they live in conical tents of birch-bark; the poles of which these are made are about 20 ft. long, united at the top; these are covered with pieces of flexible birch-bark, 'handsomely worked along the edge with horsehair thread.'

Clothing and food.—The Yakuts are capable of standing extreme cold, and are found going about in very light attire in winter, while their children sport naked in the snow. Their costume differs less in the various seasons than that of other Siberian peoples. Those who live in the Okhotsk region have partly adopted Russian costume, and partly retained the native dress—the *kukhlyanka* and *tarbass*. They also wear the *sarafan* (a long overcoat without sleeves), but not the ornamented cloak (made of fur of special cut) and the beaver cape, of which their compatriots in the Yakutsk region are so proud. Their breeches come only down the thigh, and they have long boots (*farri*) which come above the knee. Their summer-dress is the *robashka* (overshirt) and *balachon* (blouse); in winter they wear costumes of skins. The women are adept in making up fur-garments. Their dress only differs from that of the men by its greater length and ornamentation. The principal food of the Okhotsk Yakuts is dried fish; the inland tribes eat horseflesh with avidity: they prefer it to beef, and rarely kill oxen for food. One special dish of theirs is called *tar*: it is a mixture of meat, fish, various roots, grass and the under-bark of spruce, fir, or larch, pounded in a mortar; this is put into skimmed-milk mixed with water to which is added a little flour (when available), and the whole is boiled into a kind of porridge. They gather their bark harvest in spring, when the sap is rising. Their special intoxicant is *kumis* (fermented mare's milk); they also drink large bowls of melted butter. Another of their delicacies is jelly extracted from reindeer-horns and flavoured with pine-bark. They collect huge quantities of berries and cedar-cones for food. They are also great consumers of tea.

Marriage and death.—The clans consist of blood relations. Their marriages are exogamous, but the influence of their neighbours, the Yukaghir, has led the Arctic Yakuts to introduce endogamous marriage, and the frequent sexual relations that take place before marriage are always endogamous. The marriage ceremony contains an exchange of gifts, the bridegroom brings *kalym*, the bride *anna* (dowry). When the Cossacks came, they found polygamy general, owing no doubt to the nomadic character of the people, the Yakut having a wife in every pasture; but this is not much practised now, partly because the high price of the *kalym* is a deterrent, partly because girls die more frequently than boys in infancy, as they are not so carefully tended. When an important man is buried his best horse is killed and eaten, and the head and hide in one piece are set up as a memorial. In the case of a woman a cow takes the place of a horse.

Religion and superstition.—The Yakuts have been nominally Christian since the beginning of the eighteenth century, when the Tsar declared that they had merited this privilege, of which many of them remained absolutely unconscious. They go to church, but have a very hazy idea of their nominal religion. Their native chief god is Tangra. They have also many malevolent spirits, who are distinguished by the names of colours; cattle and horses are sacred to the spirit whose colour they bear.

They have many customs and superstitions. No woman other than the hostess may give anything to eat or drink to a male stranger before the fire-place, but must walk round the chimney to present it. It is wrong to wash plates; such waste will produce scarcity; earthern vessels are cleaned by being burnt; it is not felt then that anything is wasted because the food thus consumed is an offering to the fire, which also receives a small spoonful as an offering of thanks. Every Yakut has two names, by one of which he is never called.

(iv) Mongolic Tribes

Buryats

Territory.—The Buryats live about the shores of Lake Baikal in the Provinces of Irkutsk and Transbaikal. They extend from the frontier to the Lena valley and the north-east end of Lake Baikal, and from the River Ingoda to the River Oka. There are 11 tribes of them, 4 west of Baikal, 7 east.

Racial affinities and language.—They are the principal Mongol tribe in Siberia, and speak a Mongol dialect, distinct alike from the literary Mongolian language and from the speech of the Kalmuks, who are closest to them in racial affinity. This tongue they have preserved in comparative purity, as knowledge of reading and writing is common among them, and they have books of their own for which they employ the Manchu alphabet. There are three distinct dialects of it.

Divisions.—The north-west group of Buryats are known as Bargu-Buryats, the south-west group as Mongol-Buryats, and the eastern group as Aga-Buryats. The first two groups claim their descent from two different ancestors, Bukha Noyna and Bargubata. The Buryats round Novi-Selenginsk claim as their ancestor Jenghiz Khan. He it was who forced this people northwards in the thirteenth century, when they arrived on the upper Amur.

Numbers.—The population of the Buryats is increasing. The most recent estimate is 288,599, of whom 175,717 are males. About three-fifths of these live in Transbaikal and two-fifths in Irkutsk.

Relations with Russians.—The Russians conquered them in the seventeenth century after a vigorous resistance, but since then they have been among the most peaceful and trustworthy of the inhabitants of Siberia. They join the Cossack regiments of the Transbaikal, and even in 1761 there was a regiment of Selenginsk Buryats. They do not chafe at having to perform military service, which probably is not an unwelcome interlude in the life of the *yurta*.

Social institutions.—The Russians have developed their native clan-system, grouping several of their *yagans* (clans) into administrative clans. The head of the clan is called *shelenga* (elder). The group of clans is called *vyedomstva* and is presided over by a *taisha*. One feature of the collective life of the clan is the co-operative hunt, which sometimes lasts for months. Over it presides the *tubuchi*, whose office is often hereditary. Another collective act of a clan is its *tailgan*, a public sacrifice offered on behalf of the clan.

Physical appearance and characteristics.—The Buryats are broad-shouldered, inclined to be stout. They have big heads, square faces, small slanting eyes, high cheekbones far apart, broad and flat noses, low foreheads, thick lips, swarthy and yellowish complexion, jet-black hair and scanty beards. The hair is cropped very close except on the crown of the head where it is made to grow in a long queue that hangs down at the back. In temperament the Buryats are phlegmatic and patient; they lack enterprise, but have a certain amount of energy as shown in their hard work as agriculturists: it seems scarcely fair to say of them that they are only made to work by the stimulus of hunger. They are fond of drink and tobacco, and even small children smoke. They are an intellectual people, and under the influence of Lamaism have books of their own.

Occupations.—Their principal occupations are rearing cattle and horses, and they show great attachment towards their horses. They also hunt and fish, and under Russian influence cultivate to some extent rye and wheat. When they have taken to agriculture, they have proved better farmers than the Russians, being indefatigable in manuring and irrigation. They are specially adept at silversmith work, which is known throughout Siberia as *bratski* work, *bratski* (=brothers) being a name of the Russians for the Buryats. They are also successful as leather workers and in the manufacture of textile fabrics.

Dwellings.—Their *yurtas* are not in rows, but scattered; and they are surrounded by large enclosures; at a distance from the groups of their dwellings are the large enclosures (*ugugi*) where the cattle graze during winter and where huge crops of hay are obtained during the summer. They are so much attached to the life of the *yurta* that when they live in houses they make a hole in the roof and have a fire in the centre of the floor.

Clothing, food.—In summer they dress in silk and cotton, but in winter in fur and sheepskins. A wealthy bride's dowry will sometimes consist of as many as 40 cases of the richest furs. Their staple food is boiled mutton. They drink a great deal of brick-tea and blend with it rye-meal, mutton-fat, and salt. In the north they use wood as fuel, in the south camel's dung (' argols ').

Marriage and death.—Marriages are arranged among the Buryats by two families exchanging daughters; if there are only sons a *kalym* has to be paid, which consists of so many cattle. The bride receives a dowry, which counterbalances the *kalym*.

They used to burn their dead, but the practice is now forbidden by the Russian Government. The bodies of *shamans* are still sometimes burnt before being placed in the trunks of trees; if not, they are exposed on an *aranga* (platform). At the burial of a Buryat a horse (*kholgo*) is sacrificed. In former times the old people were got rid of by compelling them to swallow strips of fat.

Religion.—The nominal religion of the western Buryats is Christianity; of the eastern Buryats Lamaism, the northern form of the Buddhist religion. The Lamas (who constitute a large part of the population towards the Chinese border) are greatly reverenced; they lead ascetic lives, are forbidden the use of spirits and tobacco, do not take animal-life, and are celibates; they engage in industry outside their religious avocations. The chief of their religious houses is the *datsan* by Lake Gusinoe, where the Chambo Lama presides. A well-known feature of their religious system is the prayer wheel. The native religion is polytheistic; each class of gods seems to have a departmental head, but there is no Supreme Being.

Buryats who have married Russian wives often adopt Christianity. But Shamanism will die hard; if it has annexed the symbol of the Cross for its rites, it shows scanty signs of falling otherwise under its influence.

(v) Tungusic Tribes

Tungus

Territory and racial affinity.—The name of Tungus is applied not to one tribe, but to a whole group of tribes that extend from the Yenisei valley to the Pacific Ocean, broken up into widely-scattered groups, and, by intermarriage with their neighbours in different parts, developing very different characteristics. It is extraordinarily difficult to ascertain whether the account given by travellers refers to one particular tribe or to some widespread characteristic of the group. Thus one traveller, finding Tungus to the north and east of Yakutsk, describes them as 'perhaps the wildest, as they are the filthiest, of any Siberian tribe. They are comparatively few (at most some 4,000) and are yearly diminishing in number. They profess no religion, are nomads, and gain a living by fishing and selling furs to Russian traders, who by the aid of vodka and debauchery are slowly but surely decimating them'. He does not add a word to give any indication that there are other Tungus tribes to whom this account would be utterly inapplicable, as, for instance, the Goldi with their elaborate art. It is necessary, therefore, to be very cautious in accepting statements about the Tungus as a whole.

The Tungus are a branch of the Ural-Altaic group, and include some tribes like the Manchu and the Solons who hardly come into Russian dominions at all. They stretch from the Taimir Peninsula along the Yenisei valley, across the Vitim plateau to the sea-coast almost from Korea to Kamchatka. The Amur and Ussuri are Tungus streams. Along the Arctic Ocean not only are they found in the Nisovaya tundra, but also between the Yana and Kolima Rivers, and certain Tungus tribes, the Lamuts and Olennye, continue along the Anyuisk and Stanovoi Mountains into the Chukchee

Peninsula. The Tungus proper are chiefly in the neighbourhood of the Tunguska rivers, and between the Lower and Middle Tunguska is found one special branch of them, the Chapogir. Some of the tribes along the Amur are hybrid tribes, which are only half-Tungus, but have retained a form of Tungus speech; others have had their characteristics profoundly modified by neighbourhood and intermarriage with southern peoples.

Name of tribe.—Their diffusion over north-east Russia is probably responsible for the similarity of so many implements and customs, as they have been adopted by their neighbours. Nansen attributes to the influence of the Tungus the practical identity of type of certain things (e. g. dog-sledges) in various parts of this vast region. Their name seems to be Chinese: Tunghu = the people. The Samoyedes call them Aias (= younger brothers), which implies a late immigration.

History.—Their oldest home was Manchuria. Political upheavals in China and the conquering hordes of Jenghiz Khan drove them northwards, where earlier branches of their people (the Sucheni of history) had already gone. The Russians found them on the Yenisei at the beginning of the seventeenth century—the first mention of their name is in 1612—and in the following century they had severe struggles with their strong neighbours the Yakuts.

Divisions.—The eighteenth-century explorers divided them into Horse-Tungus, Reindeer-Tungus, Dog-Tungus—to which some have added the names of Cattle-Tungus, Taiga-Tungus, and Steppe-Tungus, the last being another name for Horse-Tungus. For purposes of Russian administration the Tungus proper have been divided into Sedentary, Nomadic, and Wandering Tungus. The former are only 1 per cent. of the whole number; they are found chiefly in Transbaikal, have intermarried with Russian settlers, and have forgotten their language. The Nomadic Tungus are cattle-breeders, who change their abode according to the season of the year, each clan having its own special region reserved for it. These constitute about 50 per cent. of the Tungus population and are found in the

Transbaikal and Yakutsk Provinces. The Wandering Tungus frequent the rivers, except in Transbaikal, and wander at large without a special region being assigned them. They pay little in taxation; they have preserved their nationality and language best. These are about 45 per cent. of the Tungus. Besides these three divisions there are about 4 per cent. who, like the Buryats, have joined the Cossack regiments. These are entirely exempt from taxation. The 3rd regiment of Transbaikal Cossacks is entirely composed of Siberian natives.

There are two great divisions of the Tungus breed. The first is the North or Siberian: to this may be assigned the Tungus proper and the Chapogir, Lamuts, and Olennye of northern Siberia, and further south two groups divided on a linguistic basis, to one of which belong the Orochon, Manegir, Birar, Kile, to the other the Olcha, Oroke, Negda, Samagir. The second is the South or Manchurian: to this are assigned two groups, on a linguistic basis, the first, with strong Mongolian influence, the Daurians and Solons, the second the Oroche, Manchu and Goldi. The physical and linguistic differences do not exactly correspond: thus the Olcha are scarcely distinguishable from the Gilyaks, who speak a totally different kind of language, while the Samagir, who belong to the same linguistic group, closely resemble the Goldi. Again, in another linguistic group we find that the Manegir and Birar resemble physically the Manchu, the Orochon the northern Tungus.

Numbers.—The population of the Tungus is, altogether, 76,507; of the Tungus proper, 62,068, of whom 31,375 are males. The numbers are diminishing, owing to epidemics and famines.

Physical appearance and characteristics.—Broadly speaking, there are two types of Tungus, which can be called North and South. The North Tungus are differentiated by their extremely short stature; they are said not to average more than 5 ft. 4 ins. They have moderately big heads with longish faces, broad at the cheeks, but narrowing to the forehead; the nose is flat and broad; the hands and feet are small. The

South Tungus are taller ; the hands and feet are distinguished by their comparative size from those of their neighbours, though a European would not like to try wearing their boots. Their heads are moderately small, with a broad square brow and an almost straight and not particularly thick nose ; the cheek-bones are more prominent than those of the North Tungus and the cheeks are hollow ; a certain amount of red shows through their olive complexion. Both breeds have black hair ; the beard is thin and short : the eyes are dark-brown and sunk. The mouth is wide, the lips thin. The Tungus have no tendency to obesity, but their figures are usually slim, wiry, and well-proportioned. The Tungus are a frank, good-natured, and hospitable people : they have been called by Castren the 'nobility of Siberia'. They are said to be very honest, and fall an easy prey to the Yakut, who trade with them. A Tungus will not receive a present unless he can give one in return, often of greater value. Keane describes them as 'cheerful under the most depressing circumstances, persevering, open-hearted, trustworthy, modest yet self-reliant, a fearless race of hunters, born amidst the gloom of their dense pine-forests, exposed from their cradle to every danger from wild beasts, cold, and hunger. Want and hardships of every kind they endure with surprising fortitude.' Despite the fact that some of them enter military service, they do not amalgamate with the Russians, who have not become assimilated to them, as they have to some extent to the Yakuts and Buryats.

Social institutions.—The organization of their family and clan system is fairly strong : it has been retained most among the nomadic tribes ; but Christianity and western civilization have had a disintegrating effect. The individual has little existence apart from his family : it is as families that they go in quest of new hunting-grounds, whereas an individual never leaves his family. A group consisting of less than 100 is called a clan (*tagaun*) ; above that number it is an *orda*. The members of a clan may not marry each other. The clan used to be governed by a *daruga* whose office was hereditary ; but in the eighteenth century the Russians appointed native administrators, whose office was mainly to collect taxes. The clan-names are usually those of ancestors, but are sometimes derived from geographical features; the Russians also give names to groups formed by themselves arbitrarily out of disintegrated clans.

Occupations.—They are not given to agriculture, but most of them are reindeer people, and the names of some of the tribes like Oroche (*Oron* = reindeer) suggests that others once were. They hunt fur-bearing animals, the most valuable of which are the black fox, which is rare (his skin is valued up to £100), and the sable—a good sable-skin fetches from 50*s.* to £10. One method of securing the sable, if he takes refuge in a hole, is to cut off his retreat with fine threads covered with bells, so that if he makes a dash out by night his movements may become known. The Tungus use the reindeer not only for drawing sledges but for riding; they use a bridle twisted round the horns, and support themselves by staffs. When they are reduced to poverty by the loss of their reindeer they live on fish. In summer the nomadic Tungus come to the coast to fish. In March, or at some appointed time, the Tungus assemble in their nomad camps to pay the *yassak.* *Yurta* and herd alike are left in the charge of the wife during the hunting season. The woman has to do all the work of the house: she has to prepare the skin for the boots, sew the clothes, skin the reindeer, and make the meals ready.

Dwellings.—They have no towns, villages, or houses as a rule, but only tents, of which there are seldom more than 2 or 3 together. In summer the *yurta* is made of birch-bark, in winter of skins or, more rarely, logs. It is constructed with perpendicular sides, and a conical roof with a vent-hole for the smoke. It is divided into sections, the best of which is taken by the owner of the tent; the others are for the grown-up children, and sometimes the labourers; but these usually have separate tents. One *yurta* will contain six or eight of these divisions. The quite poor construct their *yurtas* of fish-skin and not reindeer-skins.

Clothing, food.—Of so widely scattered a people, who dwell some of them near the Pole and others on the Chinese borders, it is impossible to describe the costume in general terms. One common costume described consists of a *parka* (blouse), *dacha* (sleeveless cloak), pantaloons worn by men and women, and cap and boots of reindeer skin. Their main food is the meat and milk of reindeer, dried fish, and a sort of cheese. They do not care for a vegetarian diet. Krahmer attributes to them the dish *tolkusha,* which has been described among the Kamchadals.

Marriage and death.—The son grows up in his father's house; he has no property of his own; the father takes even the furs that his son's skill in hunting has secured. When he is married he has his own section of the tent. The Tungus girl is free in her choice of a husband, and gets her own share of the inheritance, though less than that of her brother. Marriage is exogamous, but in the north, as with the Yakuts, we find tribes that are endogamous; so, too, there is polygamy in Yeniseisk, but few Tungus are rich enough to afford the *kalym* for more than one wife. The Tungus never burn their dead. The corpse is usually sewn up in reindeer-skin, and then sometimes put in a wooden coffin set on high posts. Among the pastoral people round Lake Baikal it is interred in the ground. On returning from the funeral these latter try to obliterate their tracks, or cut down trees to bar the way, in order that the spirit of the dead man may not pursue them.

Other customs.—The Tungus dance is peculiar; they stamp on the ground while they repeat again and again one particular word. They have few musical instruments. The influence of China on the southern Tungus has led to the development of beautiful forms of art.

Religion.—In the south a good number of the Tungus are Buddhists, but in the north Shamanism prevails. The acceptance of Christianity has led rather to a disintegration of clan customs than an improvement of faith and morals. The Tungus is generally credited with being very superstitious.

A.—Chapogir, Olennye, Lamut

Of Tungus tribes there are found in the north, the Chapogir between the Lower and Middle Tunguska, who may be a clan rather than a tribe, the Olennye between the Chaun and the Anadir, to the north of the Stanovoi Mountains, and the Lamuts who dwell in the Verkhoyansk and Kolima districts of Yakutsk, along the north shore of the Sea of Okhotsk and along the west coast of Kamchatka. They are about 2,000 in number, but are dying out. They are small and wizened, a nomadic people, who ride reindeer and drive them in sledges. They are excellent shots, and some few of them are given to fishing. They live in big conical tents, covered in winter with undressed reindeer-skin, in summer with tanned sheep-skin; these tents are easily struck, and they move rapidly. They retain their primitive simplicity and are very religious. Their Christianity, however, is tinged with Shamanism. Few among them can speak the Russian language.

B.—Orochon

In the Amur basin are a large number of tribes, who are wholly or partially Tungus. The most widely spread of these is the westernmost of them, the Orochon (or Oronchon or Orocheni). Their name implies that they are a reindeer people: we may compare Oroke, Oroche, and probably Olcha, with *l* substituted for *r*, though of these three only the Oroke still have reindeer. They dwell in the district north and east of Baikal as far as the Shilka, and along the Shilka and Amur to the confluence with the Oldoi, and northwards along the Olekma valley almost to Olekminsk. They resided originally in Yakutsk and emigrated to the Amur in 1825, occupying part of the territory of the Manegir. Those that live to the north of the Amur are called *Ninagui*, those that live to the south are called *Sholgon*. Ravenstein describes them thus: They are small and spare, with thin limbs. They have flat faces, very small and sleepy-looking eyes, of black or brown colour, noses often large and pointed, large mouths with thin lips, broad cheeks. Their hair is black and smooth ; their eyebrows are thin, and their beard is short. The Chinese tax-collectors, to whom their women were freely offered, have left their stamp on the physiognomy of many of the inhabitants. Their ordinary costume is a fur or leather frock (*gulama*) with short and wide leather drawers girt round the waist, and a belt, attached to which are a great number of things in daily use. The dress of the women is much the same, only longer. Unmarried girls can be recognized by their headbands embroidered with beads and adorned with buttons, copper coins, and small pieces of tin.

They are a nomadic tribe. Their tents are conical, easily built and removed. They have some twenty poles stuck in the ground to form a circle of from 10–14 ft. in diameter, tied together about 10 ft. above the centre. The frame is covered with birch-bark, and that again with skins of the reindeer and elk. In front there is an opening to serve as a door, and above a hole for a chimney. When they remove temporarily they leave the frame and merely remove the bark and skins. The seat of honour opposite the entrance is never occupied by women. In front are scaffoldings for drying fish and meat, and storehouses for keeping what they do not carry away on their excursions ; these are never locked, but no one thinks of robbing them.

In catching fish they use harpoons for large fish and a snare (*samolov*) for small fish.

The Orochon are nominally Christian, but keep up a good deal of the old Shamanism. They wear teeth and claws of animals as amulets, and erect idols made of wood and fur in their *yurtas*.

C.—Manegir

What is said of the Orochon is mostly applicable to their eastern neighbours, the Manegir (Manyargs, Manegr), who live along both banks of the Amur from the mouth of the Oldoi to the mouth of the Zeya, and up that river and its tributaries to about lat. 54° N. Their languages are closely akin, both accenting on the last syllable, while their more northerly Tungus neighbours accent on the penultimate, but in their physical characteristics the Manchu type is more discernible, while the Orochon more resemble the northern Tungus. The great difference between them is that, whereas both tribes are nomadic, the Orochon employ reindeer, the Manegir horses. In religion they are definitely Shamanist. They dwell on the river-banks in summer and frequent the taiga in winter. They have large herds of horses. They are said to keep their horses in condition by not feeding them for a day before a long journey, and for five or six days after returning.

D.—Birar

Further down the Amur, after a short interval, where a corner on the left bank is occupied by the Daurians, come the tribe of the Birar (called by Middendorff Bural or Byral Tungus). These dwell in the valley of the Bureya and its confluents as far as the prairie extends, i. e. up to about lat. 53° N. Down the Amur they are found to the mouth of the River Dichun. They are closely akin to the Manegir, whom they resemble in physiognomy and language, both borrowing words from the Chinese, Manchu, and Daurians. Like the Manegir, they are Horse-Tungus.

E.—Kile

East of these again, on the valleys of the Urmi and Kur, but not reaching to the Amur, are the Kile. Information about them is extremely scanty, but they seem to have been formerly a tribe of reindeer-nomads, who have settled on the River Kur and taken to fishing and hunting. Possibly the tribesman whom Middendorff calls Guragr was one of these, i. e. Kur-dweller.

F.—Negda, Sanagir, Olcha, Oroke

These tribes form a second group of Amur-Tungus, connected together by their language. The Negda (Nigidals or Neidalz (Russ.)) follow the course of the Amgun to its confluence with the Amur. They only occupy a few points beyond the Amgun. They seem to be a blend of Tungus and Gilyaks.

The Sanagir (Shanogir) are mostly along the upper and middle course of the River Gorin, but the mouth, like that of the Kur, is occupied by the Goldi, whom physically they closely resemble. They have not always been distinguished from the Negda.

The Olcha (Manguns) occupy the lower waters of the Amur from the mouth of the Gorin to Bogorodskoe and to the sea at de Castries Bay. Physically they present a Tungus type, but with great modifications; they are said by Schrenk to be a cross between the Tungus and the Gilyak, and Gilyak influence is discernible in their language; others have held them to be more definitely of Mongol origin. They are less energetic than the Gilyak, and have not to the same extent kept out the Russian traders, who have had a demoralizing influence upon them. They keep dogs in large numbers, and have a special table in their houses reserved for feeding them.

The Oroke are the last tribe in this group. They occupy settlements along the east coast and in the interior of Sakhalin. They number 749, of whom 395 are males. As early as 1709 there are allusions to reindeer-holding inhabitants of Sakhalin. Mamia Rinzo calls them Orotskoe. Klaproth thinks that the Manchus gave the name of Oron to all the Tungus people. Their character is said to be rough and unbridled. They have no permanent habitation, but dwell in *yurtas* easily removed. They own reindeer, as their name implies; a man is supposed to be well off who owns twelve. They do not shave their heads, but allow the hair to fall over the shoulders, or tie it up in a pigtail which hangs down behind. The clothes are made of fish-skin, seal-skin, and deer-skin, the latter being specially used for the trousers. The women's gowns are ornamented with brass decorations, and they have linen aprons, the material being procured by trading journeys to the Amur. For hunting they use bows and arrows and spears. Their food consists of fish, meat, roots, and herbs.

G.—Daurians

Of the south or Manchurian group the Daurians come up to the Amur below Blagovyeshchensk and occupy the right bank from that town down to Kadagan. Physically they are hard to distinguish from the Manchu; they have oval and intellectual faces; their cheeks are less broad than those of their other Tungus neighbours; the nose is rather prominent, the eyebrows are straight, the skin tawny-coloured, and the hair brown. There is a theory that they are the remains of a Chinese-Mongol military colony; but although their language contains many Chinese words, it is notable that among the names of animals those that are Chinese are the names of marketable animals, the others being Tungus. They are a tall, strong people. They have assimilated Chinese customs to a great extent. The upper classes shave their heads in front and grow pigtails behind; the lower classes do not shave their heads, and twist their pigtails round their hair. The men wear long blue coats of cotton, loose linen trousers fastened at the knee or made into leggings, Chinese shoes or boots made of skin. They have a kaftan of fish-skin or other skin, and a belt to which is attached a case that contains their most requisite accessories (knife, chopsticks, tinder, small copper pipe, and tobacco). The women dress in blue cotton gowns with short loose sleeves, above which they wear a cape or mantle of silk reaching to the waist; they carry their youngest children on their backs. They are primarily a people of huntsmen, but fish also: during the winter they secure fish on the Amur by 'malleting': the fish are visible through the ice and are stunned by a blow at the top; a hole is then made in the ice and they are secured. Their houses are set in square yards with a fence of stakes or wicker-work about them; they have a framework of wood covered with mud, the roof is covered with sedge or grass. Usually the interior is not divided; when it is, the entrance-room, where the children and domestic animals are congregated, is used as the kitchen. During the summer they have windows made of paper soaked in oil, during the winter they cover their windows with matting. Outside many of the houses are shrines containing idols with basins of incense set before them ; another religious decoration is the long pole with votive skulls adorning it.

H.—Solons

The Solons (shooters) are a tribe which are important in north Manchuria, but a mere handful of them live across the Amur. They are nomadic, and even their women hunt on horseback ; they have horses, dogs (used for hunting), sheep, oxen, and camels. Both they and the Daurians have large Mongol and even Chinese admixture.

I.—Manchu

The Manchu are the aristocracy of these tribes ; they have a proud history and have given a dynasty to China. Their real home is up the valley of the Kirin, a tributary of the Sungari. They have more marked features than any of their neighbours, thicker, more arched noses, less thin lips, bigger mouths, taller stature. They are found in the neighbourhood of Blagovyeshchensk, but there are very few beyond the Amur (the number of the whole tribe is 3,340). They are fishermen, and have boats either made of the trunk of a hollowed tree, or flat-bottomed made of planks.

J.—Oroche

The Oroche live along the coast of the Ussuri region from de Castries Bay to about lat. 44° N. Their speech resembles that of the Goldi, but we have little information about it ; the resemblances are more in vocabulary than in pronunciation. They have had their physical type modified by intercourse with the Gilyaks in the north and with the Chinese in the south : the latter have had a demoralizing effect upon them ; they have settled among them, sometimes for agriculture on a small scale, sometimes for fishing, gold-washing, and so on. The South Oroche (Tazi) have been especially affected by this intercourse. The Oroche are said to have a repulsive physiognomy. They are very short; their heads are proportionately big, their extremities small; their complexion is less dark than that of the Ainu and the hair less thick, the beard being almost nil. Their hair is black or brown, their eyebrows are strongly marked; their faces are flat and almost square; the forehead is low, round and somewhat receding, the cheekbones prominent, the eyes small and slanting, the nose small and snub, the mouth big, the lips thin. The population is 2,407, of whom 1,329 are males. Despite their name signifying 'reindeer-keepers' they are now a fisher-folk.

K.—Goldi

Territory, divisions.—The Goldi live along the Amur and its tributaries the Ussuri and the Sungari. There are three divisions of them, differentiated more by the dialect that they use than by any other peculiarity: (1) from the mouth of the Gorin where they are conterminous with the Olcha to the Gion Mountain, (2) from the Gion Mountain to the mouth of the Ussuri and up that river; (3) from the mouth of the Ussuri up the Amur and Sungari. This last group is called Kilens: the name Khodz is given to those Goldi who live below Khabarovsk.

Intercourse with Chinese. Physical appearance.—The Goldi have acted as transmitters of culture: they have absorbed much that China had to teach them and have in turn greatly affected the tribes beyond them, especially the Gilyak, who owe to them many features in their art, customs, and ideas. Their physical type has been modified by this intercourse also, and some of them present certain differences in appearance from the northern Tungus. The face is round or oval with the well-known 'Mongol' characteristics: broad cheekbones, oblique small eyes, broad, thin, low nose. But the other type is found not less frequently: there is a certain breadth in the cheeks, the eyes are less oblique, the nose is higher and more arched, the lips are thicker. Both types have black hair and eyes (occasionally grey), and have bony and muscular frames. Their beard and moustache are poor. They have one form of tattoo which is used by both sexes, viz. four spots on the forehead arranged as a cross. In character they have been described as timid, good-natured, and honest.

Numbers.—They number 5,016, of whom 2,640 are males.

Occupations, dwellings, clothing.—It is difficult to describe with great accuracy their ideas and way of life, because the travellers to whom most of our detailed information is due have not been careful to distinguish them from their neighbours. Their chief employment is fishing, for which purpose they use small birch-bark canoes for one man, and also larger boats of three principal planks, mostly of cedar-wood, fastened by wooden pegs and caulked with willow-bark, of about the length of 15 ft., adapted for crossing shallows and capable of carrying sails. They are expert oarsmen and are serviceable to the Russians in that capacity. They employ dogs to tow their boats upstream. They have an ingenious method of recovering the harpoon after they have thrown it; there is a fish-bladder attached by a line 35 ft. long and that, as it floats, indicates where it is. One method of securing fish when they descend the Amur is to fix firmly to the bed of the stream a row of tressels connected by cross-beams, and the interstices filled by wickerwork of willows: in a gap in the latter the Goldi put their fishing-nets and secure large catches. When they have dried their fish they sometimes protect them against the assault of birds by chaining an eagle in the vicinity. They do a certain amount of hunting, but have settled habitations, and are not nomads, though they may be absent from home for a long period. They are also good smiths and make beautiful ornamental spear-heads. Agriculture is confined to the cultivation of small plots of land in which they grow vegetables and tobacco surrounding their dwellings. These latter are on the lines of Chinese houses; they are built of poles with beaten clay between, or a mixture of clay and straw. The floor is covered with clay, and has a hole with charcoal in it, kept burning summer and winter, for the purpose of lighting pipes. These houses are about 30 or 40 ft. square, and will accommodate sometimes 30 or 40 people. Their costume varies, and they are most receptive of alien fashions. The same man will wear at different times a Russian overcoat, a Chinese dress or a fish-skin suit. The women are highly skilful in needlework and embroidery, and tastefully adorn their skirts and bodices.

Customs.—In a population where men are in a majority there is naturally no polygamy. They are Shamanists in religion, and bury their shamans and other great ones in huts ; the bodies of the poor are bestowed in coffins placed in trees out of the reach of wild beasts. Their favourite amusement is wrestling, and the singing of improvised songs.

CHAPTER VI

COLONIZATION OF SIBERIA

Elements of Immigrant Population—Distribution and Number of Colonists—The Present System of Colonization—The Exile System—Colonization of the various Provinces—Distribution of Russian Population—The Yellow Question and Colonization—Encouragement of Colonization in Arctic Russia.

Elements of Immigrant Population

Early Colonization : Cossacks

The colonization of Siberia has been a long process, at first gradual, and recently very rapid, that has extended over more than three hundred years. The conquests of Yermak in the reign of Ivan the Terrible were soon followed by the arrival of the first colonists. It was in 1593 that the first settlers arrived : they came from the town of Uglich, and had been too zealous in making known the plot against the Tsarevich Dmitri. But the earliest settlers were not as a rule exiles, but either traders attracted by the fur trade, or Cossacks whose settlements were extended across the continent to protect the new settlers. In 1637, Yakutsk was founded, and about the middle of the century Khabarov occupied the banks of the Amur (see Chap. XVIII). In the eighteenth century, when the borders of Asiatic Russia had become restricted, Transbaikal was largely occupied by Cossacks from the Don, and their descendants have constituted a hereditary military caste and have formed the nucleus of future military colonies in the Far East. The Cossacks have had special privileges as settlers : in central Siberia they are granted 60 acres of land per man. But it is difficult for them to combine their functions of settlers and soldiers, and the individual Cossack is usually so much a worse colonist as he has discharged his military duties effectively. (For the origin of the Cossacks see p. 357.)

In the eighteenth century a line of forts was constructed along the River Irtish, from Omsk in a south-easterly direction as a protection against the wild tribes of the Kirghiz steppes: the military occupation of this district preceded the advent of civilian settlers, which was not really developed till near the end of the nineteenth century.

Exiles

But another and very different source of immigrants soon began to be drawn upon. The first recorded mention of exile to Siberia in any Russian legislation is in a law of the Tsar Alexis Mikhailovich in 1648. Exile was largely used at first as a means of getting rid of disabled criminals, men on whom some savage sentence of mutilation had been carried out, and, as was natural to expect, such men were useless as colonists. Then at the close of the seventeenth century it was regarded as desirable to send exiles to populate new territory: an extensive criminal code supplied large contingents when exile was the recognized penalty for such different offences as fortune-telling, snuff-taking, driving with reins and setting fire to property accidentally. The discovery of mineral wealth added a fresh incentive: rich mines were found at Yekaterinburg, and this together with the establishment of manufactories in Irkutsk led to a large demand for labour, which was met by extensions of the punishment of exile to fresh crimes. In the year 1753 capital punishment was abolished in Russia, and its place was taken by perpetual banishment to Siberia with hard labour.

The exiles dispatched to Siberia fell under three designations: they were either criminal, political, or religious. In 1900 exile to Siberia was abolished, though in 1904 it was restored for political offences and the number of political exiles was greatly swelled by the revolutionary outbreaks of the period of 'Vladimir's day'. In 1906 45,000 political exiles entered Siberia. For a long time the exiles (*kolodniks*) were driven in herds from one village to another without any proper arrangements being made for them, and they were often starved on the way. At the beginning of the nineteenth century more satisfactory arrangements were made : in 1811 a suitable force of regular guards was organized to convoy parties of exiles, and all exiles were furnished with identifying documents (*statéini spiski*) to show who they were and whither they were bound. In 1817 étapes (*ostrogs*) were established along the main roads; they stand now easily identified in the villages in western Siberia, often, it is said, the most cheerful-looking building in the community. In eastern Siberia they are usually outside. In 1823 a bureau of exile-administration was established in Tobolsk, which has since been removed to Tyumen, and through that has passed the endless procession of exiles, political and criminal. No record of them was kept until 1823 : since that date to 1898 there have passed by 700,000 exiles ; and with them 216,000 voluntary followers (*dobrovolni*). The criminal prisoners fall under two heads : (1) *katorgeni rabotniki* (criminal convicts), who are sentenced to hard labour ; (2) *poselëntsi*, who are condemned to shorter periods of imprisonment. During this same period 187,000 criminal exiles with 107,000 companions have entered Siberia.

Since 1870 the government has done its best to keep apart the criminal from the peaceful colonists. From that date the convicts were confined more and more to the Lena territory and parts of the Far East, while until recently the northern part of the island of Sakhalin retained by Russia has been almost exclusively a penal colony. The political exiles have been mainly in west and central Siberia. A great terror to life has been added by the escaped convicts (*brodyagi*) in various parts of Siberia. Escape is easy ; one writer in 1902 says that one-third of the transported escape all control and wander, rob, and terrorize. On the other hand the political exile who settles in the country often adds the most enlightened element of the populace to the community : such men settle in the towns ; they open shops and enter into the social life, being readily trusted and hospitably treated by the inhabitants, and they found a progressive public opinion. They have been among the best sections of the community, and have added much to the development of the country : some of them have been among the most enterprising of its explorers : e. g. Bogoras, whose work on the inhabitants of the Chukchee Peninsula is the most important contribution to our knowledge of that district. Some political exiles have become the trusted agents of the government in the districts to which they have been sent.

The exiles who were sent for religious reasons need a special word. The great occasion for their banishment arose from the opposition to the reforms of Nikon in ecclesiastical matters, and afterwards to those of Peter the Great in social and political affairs. Nikon in the middle of the seventeenth century had tried to make the ritual of the Russian Church conform more exactly in certain small matters with that of other branches of the Eastern Church, removing certain errors from the liturgical books. To a conservative opposition this was apostasy, and when Peter came, Nikon was Antichrist. So there arose great numbers of dissenters (*raskolniki*) : their first great division was into Popovists or supporters of the priesthood, and Bezpopovists who felt that the priesthood was hopelessly vicious, as the Church had become apostate. Both parties believed in Apostolic Succession, and the Popovists had carried with them in their secession a bishop through whom the succession from the apostles might be transmitted ; but he died before he could consecrate a successor. There was even some talk of cutting off his hands that their 'laying on' might convey the gifts of the Spirit, but, as the words of consecration could not be said, the idea was abandoned. They had therefore to depend upon runaway priests. The Bezpopovists resorted to other and more direct methods of a continuance of spiritual grace : e. g. they would stand looking upwards with their mouths wide open, that spiritual blessings might reach them from heaven by that means. These *raskolniki* are ascetic, industrious, and abstemious, abstaining entirely from alcohol and tobacco. They are the predominant element among the settlers of Transbaikal, at any rate among those who have been settled for any appreciable period upon the land : the Don Cossacks are nearly all *raskolniki*. They are not a learned folk, but at the same time they encourage elementary education sufficient for reading the Bible.

There are other sects, some of which indulge in extravagances that contravene all morality. Among the more peculiar sects are the Skoptsi, found chiefly in settlements round Yakutsk.

In dealing with religious dissent periods of persecution and toleration have alternated, but the *raskolniki* now constitute as much as 10 per cent. of the population of the whole Russian Empire.

Enforced Settlement on lines of Communication

Besides the actual exiles another class of settlers should be mentioned, viz. those peasants who have been bound to settle at appointed places in order to maintain communications along the roads, and the *yamshchiks* who drive travellers along the various sections of the post-roads.

Free settlers

The last element in the immigrant population of Siberia and much the most important is that of the free settlers, whether helped by the government or not. Voluntary immigrants (*samovolni*) settled of their own accord in the remotest parts of Siberia, and founded communal colonies as in European Russia. Being quite cut off from their kindred by reason of the immense distance, they intermarried with the Cossacks or with the natives. Where the Russian element has not been very strong they have even adopted the language of the natives among whom they dwell. Thus the Russians who live among the Buryats and Yakuts have often adopted their language. The descendants of these settlers are the Siberiaks who have lost their national sentiment while retaining many of the habits and customs of Russians. This resulting race is shorter and of darker complexion, of lower morals but stronger independence. During the eighteenth century and for a considerable portion of the nineteenth the Russian government did what it could to discourage the samovolni.

But of recent years the warmest encouragement has been given to free settlers in Siberia by the Russian government. This was made possible by the abolition of serfdom in 1861; not till then could the Russian peasant leave his native soil. For the first 300 years of Russian occupation something like 3,000,000 persons immigrated from European to Asiatic Russia. But there was a marked increase in the number of immigrants in 1896, and since the Russo-Japanese War more have arrived than the numbers during the three previous centuries. In 1914 the sum of £3,000,000 a year was voted for the development of colonization in Siberia, whereas previous to 1896 only £100,000 had been voted and in 1906 £500,000. Between 1909 and 1913 the area of new land surveyed and parcelled out for colonization was 18,000,000 desyatins (i. e. 75,850 square miles), and 350,000 families were settled. To this have to be added 6,000,000 desyatins (25,000 square miles) of land belonging to old *mirs* (village communities) brought under cultivation during those same years.

Distribution and Number of Colonists

Colonization has been mainly directed to the black earth zone, through which run the Trans-Siberian Railway and the great Siberian road. It extends from about lat. 58° 30′ N. to 55° N., though in the Altai district the region of colonization comes as far south as lat. 51° N., where there has been a great inrush of colonists since the 'Cabinet' estates of about 200,000 square miles were thrown open for general colonization in 1906; between 1896 and 1909 one-third of the 3,000,000 immigrants to Asiatic Russia went to the Altai district. In the same period about 900,000 went to the Steppe region, about 258,000 to the Tobolsk region, and about 300,000 to the east of Lake Baikal. The Tomsk Government (outside the

Altai) attracted nearly 500,000 and the Yeniseisk Government nearly 400,000. Only 18,000 went to Turkestan.

The growth of colonization can be shown by the following figures. From 1870 to 1890, 500,000 settlers entered Siberia; from 1896 to 1905, 1,078,000. The years that followed the Russo-Japanese War have shown even greater numbers of colonists, as the following table shows :—

Year.	*Settlers entering Asiatic Russia*
1906	141,294
1907	427,339
1908	664,777
1909	619,320
1910	316,163
1911	189,791
1912	201,027
1913	234,877

During the five years between 1909 and 1913, 75,850 square miles were parcelled out for colonization and settled by a new population of about 1,500,000. During this period also about 6,300 miles of road were constructed. The decrease in the number of emigrants in 1910 was attributed to the first of a succession of good harvests in southern Russia.

There is a marked decline in the number of emigrants who return each year to Russia. Excluding *khodoki* (see below) and migratory labourers, the percentage of returning colonists has fallen to about 4 per cent. a year. This applies to all the districts except the provinces of Yeniseisk and Irkutsk, and in those two colonies special climatic conditions had reduced many colonists to destitution.

The Present System of Colonization

Selection and tenure of land

The colonization of Asiatic lands is directed by the Ministry of Agriculture. Every effort is made to secure that immigration shall be popular, and to encourage the development of the remoter parts. By now the more attractive parts for settlement are practically all occupied, including the districts through which the railway runs and the imperial lands in the Altai district, but settlers can obtain loans from the State if their land is in the taiga where much clearing has to be done, or in the steppes where deep wells have to be dug, or in the far east on account of its great distance. The loan for difficult regions or lands in the far east may be 400 roubles, and in Transbaikal 250–400 roubles. These loans are on easy terms: nothing at all is paid for the first three years, and after that the debt is paid off in instalments for the next 10 years. Sometimes the second half of the loan is remitted altogether. One condition, however, is always enforced, and that is that a *khodok* goes out in advance to view the land. A *khodok* is a man who is sent out to see whether the land is suitable for colonization. One *khodok* may represent several families, but not more than five. He usually goes a year before the intending settler, and after selecting the land returns to Russia. He must travel with a certificate, which is given gratuitously. From April to June is the best season for judging the land, but the settlers themselves go out before the winter is over for fear of losing their chance of securing the land that they wish: sometimes they commence their occupation before the snow is gone, which precipitancy may lead to very disappointing results for them. *Khodoki* are not included in the table given above.

Privileges of emigrants

Other privileges are conferred on settlers. For the first five years they are exempted from taxation, and for the next five years they only pay a half. Settlers above the age of eighteen have their military service postponed for three years. In the eastern parts of the Amur General Government and in Turkestan settlers over fifteen have six years' postponement of service. Such Russian colonists as inhabit the lower valleys of the Lena and Yenisei are altogether exempted from military service as an inducement to continue their struggle against the forces of nature. Should the elder son die while the younger is in the ranks, the latter is immediately given his discharge and goes back to the farm. Further, the government has started depôts for agricultural machinery of modern type. In 1909 there were 64 of these depôts, in 1913 there were 300. In 1898 the value of the agricultural machinery imported into Siberia was 211,900 roubles ; in 1913 it was 8,400,000.

Again, in his first year the new colonist is supplied by the government with enough seed-corn to enable him to sow three desyatins of land. Along the routes there are hospitals for the immigrants, who travel in fourth-class carriages (those carriages marked as accommodating 40 men or 8 horses) at a cost of 3 roubles for 1,000 versts, which means less than 1*s.* for 100 miles. Children under 10 are carried free. Baggage, horses and cattle are taken at very low rates. There are thirty stations arranged for the distribution of land, the largest of which is at Chelyabinsk.

At Chelyabinsk, Kansk, and Stryetensk a large number of houses have been erected for the temporary accommodation of the immigrants. Along the lines are stations where free medical aid is given. Hot food is served out free of charge to children under 10 and sick people. Those who fall victims to infectious diseases are given free treatment in government hospitals.

Ninety-six per cent. of the land in Siberia belongs to the State : only in the Amur territory is it ever purchased. The usual amount allotted is now from 8 to 15 desyatins to each male member of a family. The Cossacks have had land allotted to them on a more liberal scale, each male having been given in central Siberia 60 acres, in the Amur General Government 100. New land for 'freehold' farms is allotted in parcels of 25–50 desyatins of arable land to each family irrespective of the number of males. An immigrant from Russia gets on an average a farm nine times as large as that which he had in Europe. The land is conveyed to the settlers by letters of allotment : it continues to be state-property, but for the perpetual benefit of the settler, who has no right to sell or mortgage it.

All the land set apart for communes of immigrants is now carefully surveyed, parcelled out and assessed for loans and taxation.

Conditions of colonization and sources of emigration

The great incentive to colonization from Russia is that the peasants in Europe occupy plots which only require work for from 60 to 75 days in the year, and cannot find work in the neighbourhood. In Asia they are assigned allotments large enough to give them work for the whole year. They come mainly from the northern part of the black earth zone in middle Russia, and to a less extent from Lithuania and the Governments of Perm and Vyatka. They are apt to retain the local characteristics of their old home ; thus the settlers from the Government of Poltava tend to dislike innovations, and those from the Government of Mogilev do not generally bear a good name as efficient colonists. Every year the Russian government points out what tracts of country are open to colonization, where villages may be built, what improvements have been made in irrigation. The colonial villages are definitely planned and scheduled before they contain a single inhabitant.

The sites of villages are often determined by strategical considerations, the military authorities requiring a certain number along the line of this or that hill or valley. The class of Russians who are allowed to settle in Asia is determined by the government. Thus in 1914 the government declared that the following might settle beyond the Ural : ‘ all peasants and those engaged exclusively in agriculture, and also artisans, workmen, factory-hands, merchants and shopkeepers. People of other classes must, before emigrating, apply to the governor of the province in which they live.’ It further publishes a summary of information showing intending colonists exactly what their status and privileges will be. Those who belong to forms of religion that discountenance military service are not allowed to settle in certain provinces (e. g. Semiryechensk).

The Exile System

Improvement in Condition of Exiles

To the outer world Siberia has been more associated with the exile system than with any other fact, and novelists, journalists, and travellers have made its features familiar; but the long march of many months along the roads has given place to the more rapid transit by the *arestantski* wagons on the railway, and the abolition of exile for ordinary criminals has completely transformed its associations. For some years there have been no processions of convicts, the *katorgeni* shaved on the right side of their head, the *poseléntsi* shaved on the left, nor daily tramps of 20 miles from étape to étape, with military guards relieving each other at the intermediate stages (*polétape*).

Relations of Exile with Colonization

The effect of the exile system on colonization has been bad. For a long time the exiles were not sent to unoccupied land, when their period of detention was over, but were attached to existing settlements, though not exceeding a proportion of one-fifth of the older inhabitants. After that an effort was made to detach the actual convicts: they were sent farther away, while the political exile, who had usually not been banished to so remote a region, took part in the urban life. The political exiles are of two kinds, those who have been sentenced after a legal trial, and those who have been banished by the more arbitrary system of administrative order from the Minister of the Interior. These two classes of prisoners are divided into those with rights and those without. The former can occupy land and earn wages; the latter are restricted to a certain number of small trades and their annual turnover is limited. They are employed generally on wharfs, railways, &c., and in remote districts on post-roads at the current rate of wages. A certain amount of convict-labour was also employed on railways, with some success in central Siberia, but without success on the Ussuri Railway.

Eight months of work on the railway were allowed to count as a year of imprisonment. Gangs of escaped convicts (*brodyagi*) have always been a terror to remote regions. These *brodyagi* find it easy to get away, and the inhabitants leave out food for them, not so much from charity as from policy, that they may be discouraged from taking it forcibly. If any of the *brodyagi* are found dead with wounds in front no particular questions are likely to be asked about their death, as it will be assumed that they have been killed by somebody in self-defence. It is only when they are found dead with their wounds behind that investigation is made.

The scattering of the prisoners through Siberia was less a wrong to them than to the country. Their actual treatment seems to have compared favourably with the treatment of criminals in most countries ; but the permanent element that survived in the country, apart from the political prisoners, did not make for its welfare. For every fifty-seven inhabitants Siberia has received one criminal or political prisoner, and in 1898 (two years before the abolition of the penalty of exile) the number of transported persons was 298,574. It must be remembered also that a large proportion of the exiles (from 1867 to 1876, more than 50 per cent.) had been banished because they had been found refractory or otherwise undesirable by their *mirs* at home, so that the land was being filled by drafts of those who would presumably make the worst settlers. Transportation again does not contribute to colonization owing to the large proportion of unmarried persons. But, worst of all, the country had a bad name, and the development of its great resources was thwarted by its penal associations which acted as a deterrent to those who would have made its best colonists.

Colonization of the various Provinces

Tobolsk.—This was the province in which colonization was first encouraged, when colonists were settled along the banks of the Rivers Tura, Tavda, Tobol, and Ob. It is the province, too, in which are the largest number of political exiles. In

1910 a writer speaks of them as numbering 40,000. Among these are a very large contingent of Poles, many of them the descendants of former exiles : there were many Poles banished in 1758, 1831, and 1863. For a long time Tyumen was the great centre of organization of settlers on the land. Now these functions are mainly performed at Chelyabinsk ; there the settlers are arranged into parties and sent under superintendence to the district which they are to colonize. They have to wait for formal permission to settle, but this precaution is merely taken in order to prevent debtors from absconding. In May 1896, when the great rush of immigrants to Siberia began, there would be as many colonists on a particular day passing through Chelyabinsk as the whole population of the town (viz. 17,000). Since the opening of the steppes the rush for Tobolsk has considerably diminished, but the population is still decidedly increasing. In 1858 it was 1,021,266 ; in 1897 it was 1,438,484 ; in 1911 it was estimated at about 1,975,239. Ninety-three per cent, of the population is Russian : the rest of the population includes about 40,000 Tartars, Bokharians, and Kirghiz, 20,000 Ostyaks, and 15,000 Samoyedes and Voguls. The bulk of the population is in the south of the province. Seventy per cent. live in the steppe districts of Kurgan, Ishim, Tyukalinsk, and Yalutorovsk. In the Tyukalinsk district there have been since 1802 colonies of Finns, descendants of prisoners taken by the Russians in the war against Charles XII of Sweden.

Tomsk has seen a more rapid increase of its population than any other province, the extension of colonization being especially due to the opening of the Cabinet estates in the Altai to general colonization. The northern part of the province is Crown land under the Ministry of Agriculture, but the southern districts (Barnaul, Biisk, and Zmyeinogorsk), which are Imperial lands, are the most thickly populated. Since 1865 there has been a constant increase of immigration into these regions ; in the last decade of last century, 300,000 settlers arrived, and so attractive are the prospects that this rich mining district presents that no facilities are now given to new arrivals by the government. The Altai district has been kept clear of exiles. By 1900 practically all the available land for colonization along the Siberian Railway had been taken up. But land has also been allotted in the taigas of Tomsk, Chulim, and Mariinsk, and the black earth of the taiga is extremely good, when the ground is cleared. The increase of population has been phenomenal: in 1858 it was 694,651; in 1897 it was 1,929,092; it was estimated in 1911 at 3,673,746. Of this population 93½ per cent. is Russian; one city, Kainsk, is predominantly Jewish. Other elements in the population are Ostyaks in the north, Samoyedes (about 6,500), Tartars and Bokharians ('Kalmuks'), and Teleuts or Telengites in the Altai.

Steppe General Government.—The Russians first arrived in these parts in the sixteenth century. The type has become much transformed by cross-breeding, and has grown to approximate to that of the Siberian steppe-dweller. Military occupation long preceded the coming of peaceful colonists. Their settlements along the River Irtish and along the Biisk line in the Tomsk Government guarded the country against Tartar invasion: as many as 5,174,949 desyatins were occupied by troops; but it was felt towards the end of last century that the Cossacks did little for the civilization of the wild nomadic tribes. In 1868 there was not yet a single peasant settlement. In 1875 a cry was raised for peasant immigrants, when the Governor-General of 'the Steppes stated that the civilizing effects of the Cossack settlements were very indifferent, and accordingly a survey of the Akmolinsk steppe was ordered. The Russification of the Kirghiz on the steppes has been due much more to methods of peaceful penetration by traders and settlers than to the military occupation of the country, though in Akmolinsk the proportion of Cossacks to the whole population was 109 to the thousand, and in Semipalatinsk 42 to the thousand. In introducing the peasant settlers every effort was made that they should encroach as little as possible on the rights of the native tribes. As in all provinces where there are many Cossacks, *raskolniki* abound, and in some parts

(e. g. round Kokchetav and Petropavlovsk) a certain number of Cossacks, estimated at 1,700, are Mohammedans. The greatest element in the steppe population is now the peasants: in Akmolinsk they are 54·2 per cent., though in Uralsk they are only 16·5 per cent. The Cossacks are most in Uralsk, where they are 75 per cent., while the artisans, gentry, and official class are most in Turgai (48 per cent.) and Semipalatinsk (41·6 per cent.). The Kirghiz are still considerably the largest section of the population. The population has grown enormously in Akmolinsk, less in Semipalatinsk, the figures being in Akmolinsk in 1858, 277,451, in 1897, 678,957, and in 1911 (estimated), 1,443,721; in Semipalatinsk in 1858, 217,451, in 1897, 685,197, and in 1911 (estimated), 873,760.

Yeniseisk.—The bulk of the land in this province lying to the north is stony and swampy and unfit for cultivation. The district of Turukhansk, which comprises more than two-thirds of the territory, only contains a small proportion of the population. Fully 80 per cent. of the inhabitants live in the district of the railway. The most favoured part of all is Minusinsk, where, attracted by the best climate in Siberia, a large number of settlers have planted themselves at their own risk. In 1907 the Government Survey Staff surveyed and prepared for incoming colonists unoccupied territory to the extent of 102,600 desyatins. When the Russians settled in this district they drove the native inhabitants either north to the tundras or south to the Minusinsk steppes. The native inhabitants include in the north, Ostyaks, Samoyedes, Tungus, Yuraks, and Yakuts; in the south, 'Kalmuks', Teleut Tartars, Chern Tartars, Sagaïs and Abakansk Tartars. In the Achinsk and Minusinsk districts the native population was growing in 1900. The Russians, however, are about 90 per cent. of the population. Yeniseisk had 303,256 inhabitants in 1858, 559,902 in 1897, and in 1911 they were estimated at 966,409.

Irkutsk.—The conditions in this province have been much the same as in Yeniseisk. In both, the bulk of the population is settled along the railway. But more than in Yeniseisk the district was used as a dumping-ground for convicts. Before 1900, convicts to the number of 950 to 1,000 were forwarded annually. Until the Siberian Railway was opened there was very little immigration into this region, but it is now rapidly developing. In 1896 grants of 15 desyatins per man were made. Native populations, which a few years ago were reckoned at 21 per cent. of the whole, include Buryats, Tungus, Tartars, Ostyaks, and Soyots. In 1858 the population was 222,533, in 1897 it was 506,517, and in 1911 it was estimated at 750,000.

Yakutsk.—There is little that can be called colonization in this vast territory. Russian population is only found where there are mines: even the posthouses along the Lena post-road are largely kept by Yakuts, and where the Russian population is sparse it has adopted the Yakut language. In 1897 the population was 261,731; in 1911 it was estimated at 277,187.

Transbaikal.—The conditions of this territory are somewhat special. Peasant colonies are rare; most of the settlers pass through it and fix their habitations in the Amur district. In 1900 it was estimated that only 12 per cent. of the settled land was occupied by peasants, while 45 per cent. belonged to the natives and 35 per cent. was in the hands of the Cossacks. But only 27 per cent. of the total area of the Transbaikal is cultivated at all, 40,000,000 desyatins lying waste. There are a great number of exile-settlers in this district, the abode of many who have become vagrants not being known. The most famous exiles of this district were the Dekabrists, as those who took part in the plot of December 14, 1825, are styled. These were made to construct their own prison in Chita, and by their improvements in draining and levelling transformed the place from a village into a prosperous settlement, now the capital of the province. The Cossack occupation of the province dates from the middle of the seventeenth century; in the eighteenth century it was more fully enlarged and organized. When in the fifties of last century the Amur region was absorbed by Russia, Transbaikal Cossacks were transferred there. The land assigned to the Cossacks amounts to 3,000,000 desyatins: it is under the management of the community represented by the *stanitsa* or village and the *sotnya* or group of a hundred soldiers. A great number of *raskolniki* live in this territory. It was the scene of exile of the famous dissenter Avakum. The Russian element in the population in 1900 was estimated at 64 per cent. The peasants are settled principally in the Selenginsk, Verkhne-Udinsk and Chita districts; the Cossacks occupy the land near the frontier and the villages of the Rivers Dzhida, the lower Chikoi, the Onon, the Ingoda, the Shilka, and all the eastern portion of the territory. Sometimes their settlements alternate with peasant villages. The native tribes are the Tungus and the Buryats: the former are chiefly in the districts of Chita, Selenginsk, and Barguzin; the latter mostly in the same districts and Verkhne-Udinsk. The growth of population has been steady: in 1858 it was 352,534; in 1897 it was 664,071; in 1911 it was estimated at 868,790.

Amur.—The beginning of Russia's colonization of the Amur territory was in the spring of 1857. A regiment of three *sotnyas* of Transbaikal Cossacks was ordered to settle with their wives and children along the Amur. They came down the river on rafts and were settled in *stanitsas* along the river at distances varying from 12 to 18 miles, the distance being determined rather from the desire of keeping up communications than the suitability of the places for agriculture. Their task was to defend the frontier towards China and to provide postal communication between the Amur district and Transbaikal. The original settlers suffered terribly from the natural difficulties with which they had to contend, so that, although it was a peaceful occupation, the casualties were equivalent to those of a campaign; later, in 1877, an inundation spoilt most of the land assigned to the Amur Cossacks. In four years from their first arrival they had established throughout the Amur basin 60 villages with a population of 11,850. The land had been previously uninhabited with the exception of the confluence of the Zeya and Amur, where Chinese Manchus lived.

In 1869 peasant colonization began, the first settlers being religious sectaries from the Tauric and Samara Governments. There are a large number of *raskolniki* in the territory, estimated at 10 per cent. of the population, the most prominent being the Bezpopovists and Molokani (milk-drinkers). Immigration is now on a large scale, even where no government assistance is given, and despite the journey of a month or six weeks required to reach this region. Whole families are added to communes which still have free land at their disposal. Colonists, where there are fifteen or more families from the same place, form a commune together, which receives its name according to the desire of the settlers. Nansen gives the figures of the total population in 1911 as 286,263, of whom 43,959 were non-Russians: that is to say, mostly yellow men. At Blagoslovennoe, at the junction of the Samara and Amur, there is a settlement of about 1,000 Koreans. Manchu-Chinese are found mainly along the river for 44 miles below Blagoslovennoe and for 14 miles inland. In 1897 the population of the territory was 118,570. In 1911 it was estimated at 286,263.

Maritime Province (Primorsk).—Following on the occupation of the Amur territory came the occupation of the Ussuri district. Battalions of Cossacks from the Transbaikal had orders to settle along the Ussuri: the colonization of the district began in 1859, but progressed slowly. As in the Amur territory, the original settlers suffered terribly, and were assigned land without proper care being taken to see whether it was suitable for colonization. The actual journey to their new settlements took a year and a half. A big inundation of the Ussuri added to their troubles, and their morale was not improved when they were joined by an army which had been sent there for a punishment. They had difficulty in resisting a Chinese revolt in 1868, and many of them became dependent on the Manses (Manzi) for money help and found them rigorous and exacting masters. In 1882 an experiment was made; a three years' trial was to be given to 250 families brought annually from Odessa at the cost of the Government,

315,000 roubles per annum being assigned for the purpose. Altogether to the Ussuri district there have migrated between 1859 and 1913 about 250,000 Russian peasants, exclusive of the Cossack population : some of the new settlers came from Siberia, some from European Russia. The Cossacks were posted along the Ussuri valley, but further investigation was made to see how much room there was for more settlers, and in 1911 the colonists were granted Cossack land to a reasonable extent; the Cossacks, who had the pick of the land, not being as a rule very good settlers. In another part of the province, sailors stationed at the mouth of the Amur have been allowed to retire after 15 years' service, have received a plot of land, and have been permitted to send for their wives and children at the expense of the Government. The Russian population is settled mainly in the valleys of the Rivers Suifun, Lefu, and Suchan, about Lake Khanka, along the right bank of the Sungacha and Ussuri, and in the district about Olgi Bay. The great increase of immigration after the Russo-Japanese war found the local authorities unprepared, and the defective arrangements had a prejudicial effect in later years. Nansen gives the figures of the total population in 1911 as 523,840, of whom 360,437 were Russians; but these figures do not include the Chinese and Koreans, who work during the summer and then return to their homes. The number of settlers from 1900 to 1909 was 142,674, and in the year 1913, 13,011.

Kamchatka is the name for the territory detached from the Maritime Province, including all that is north of lat. 56° N. It is not at all adapted for colonization. The attempts made in past years have been practically abandoned. Only a few Russians are found at isolated points, and the entire population in 1911 was only estimated at 36,012. Investigations into the possibility of colonizing Kamchatka are proceeding.

Sakhalin.—By the Convention of March 18, 1867, Russia and Japan secured the common right to occupy unappropriated places all over the island. This caused a keen competition between them, but Japan had the advantage of being a close neighbour, and the Russians were forced either to draw settlers from Europe by the inducement of great privileges, or to found settlements of unmarried soldiers, a system which was valueless for the purpose of peopling the country. The Russians had not enough men to occupy the desirable places in the island ; so they erected posts which bore inscriptions to show that occupation had taken place. This method was promptly copied by the Japanese. In 1869 a party of 800 convicts was sent to Sakhalin, and when it became a penal colony all women sentenced to hard labour were forwarded to this island from European Russia 'with a view to secure the family principle'. All exile settlers receive grants of land and a loan from the Government for the organization of the household ; on obtaining a good character they are allowed to settle in the Amur and Maritime territories. On April 25, 1875, the Japanese share in Sakhalin was exchanged for the Kuril Islands, but by the treaty of Portsmouth (1905) Russia lost all the island south of lat. 50° N. In 1897, while the island was entirely in Russian hands, the population was estimated as follows :

Official class (military and civil)	2,500
Peasants	8,000
Exile settlers	7,500
Exile convicts	7,000

The population of what remains to Russia was estimated in 1911 as 8,849.

The colonization of the Far East has been a matter of the first political importance for the protection of the Russian interests, but a great deal has to be done to make it a success in view of the special difficulties. The loans, usually of 150 to 200 roubles, are too small ; there is a great dearth of roads and no sufficient organization of development in road-making ; there are no possibilities of a large sale of the produce of the farms in the neighbourhood ; greater efforts should be made to secure that the new conditions shall be approximated as much as possible to the old conditions of the settlers.

Distribution of Russian Population in Siberia

The growth of population in Siberia, mainly due to immigration, has been very remarkable. A population which at the end of the eighteenth century was about 1,500,000 is now estimated at more than 12,000,000, including districts of Central Asia. Even in the forty years between 1858 and 1897, before the great wave of immigration, it had doubled itself. The same rapid growth may be seen in particular towns; as one illustration Novo-Nikolaevsk may be mentioned. It was founded in 1896, and in 1913 it had a population of 70,600 inhabitants.

The areas inhabited by Russians are the following:

In the north along the great rivers we find Russians in isolated districts along the Ob: e. g. Narim, the confluence with the Irtish, Berezov, &c. The Yenisei valley and the Lena valley down to Yakutsk are settled by Russians and so are lateral strips of land between, along the river Vilyui and to the Lower Tunguska; further east there are isolated points, such as Verkhoyansk and Nizhne-Kolimsk.

The black earth belt is the chief zone of Russian occupation: the northern boundary of settlement runs by Verkhoture, Turinsk, Tobolsk, Tara, Kainsk, Tomsk, Yeniseisk, thence along the line of the Upper Tunguska to the upper Lena; the southern boundary runs south of the great Siberian road and Siberian railway from Verkhne-Uralsk to Omsk; thence along the River Irtish till it nears the Mongolian frontier. The boundary line of the Russian district then runs north almost to Mariinsk; then with a détour to include the district round Minusinsk, the line runs south of the great Siberian road to the south-west corner of Lake Baikal and along the eastern shore of the lake.

There are 'islands' of Russian occupation about the Kirghiz steppes, e. g. Kokchetav, Atbasar and Karkaralinsk.

Other districts in Eastern Siberia occupied by Russians are:

(1) The Selenga valley to Kyakhta.

(2) The region from Chita to the confluence of the Rivers

Shilka and Argun, including all the district between the rivers from a line drawn SW. from Chita to the Mongolian frontier.

(3) The left bank of the Amur to Khabarovsk.

(4) Both banks of the Amur from Khabarovsk to Nikolaevsk.

(5) The Ussuri valley and the district between Vladivostok and Olgi Bay.

(6) Okhotsk.

(7) Certain districts in Kamchatka.

The Yellow Question and Colonization

Advantages and disadvantages of Chinese labour

A Russian in the eastern provinces would find it difficult to say whether it was hardest to get on with or without the labour of the yellow races. On the one hand the presence of workers who are frugal and content with low wages, of great skill and application, is likely to be a formidable menace to the position of the Russian immigrants, and the immigration of a yellow population, all of them able-bodied, so that soon there will be one oriental labourer to every able-bodied Russian, seems likely to justify the warning of Li Hung Chang that Russian interference with China would turn Siberia into a Chinese province. On the other hand there is no doubt that the resources of the land have been enormously developed by Chinese labour ; we have only to compare the appearance of Vladivostok, where Chinese labour has been largely employed, with that of some purely Siberian town to realize how dependent these eastern provinces are for their prosperity on the yellow man. In Vladivostok, Russian paving had to be taken up almost as soon as laid and replaced with the work of Chinese labourers ; a quay erected by Chinamen soon replaced one clumsily erected by Russians. Russia has to face the problem whether her eastern provinces shall be developed efficiently by the labour of aliens who will surpass her own population in numbers and resources, or whether the yellow races shall be excluded and the provinces be less completely developed. Already, in 1904, of 487 industrial undertakings 192 belonged to Orientals and employed no Russians, while 295 belonged to Russians but employed yellow labour. In the villages the Chinese secure a monopoly of trade among the local population, supplying them with all necessaries. At present the policy is directed towards exclusion, and it is only surreptitiously that Chinese labour can be employed in the Amur and Ussuri valleys; at least that was so just before the present war broke out, but with the able-bodied men called up for military service it is likely that the rigour of the policy initiated by the Governor-General of the Priamur will have to be considerably modified.

Strength and distribution of Chinese

In the Ussuri district the original masters of the land were the Chinese tribes of Manses (Manzi), and they have to a considerable extent reinstated themselves in their old position, by the money help that they have given (at usurious interest) to the Cossack and peasant settlers and by their bullying methods of trading with the native inhabitants. They are in a very strong position, as under the terms of the Treaty of Peking (1860) Chinese offenders on Russian soil have to be tried by Chinese magistrates even for minor offences. In the interior of this same region the Chinese have villages of their own which are governed by their own headmen. The ruthless treatment of the Chinese at Blagovyeshchensk in 1900, when numbers of them were driven into the river and drowned there, checked for a time Chinese settlement along the Amur, but the shortage of agricultural labour during the Russo-Japanese War gave an opportunity to the Chinese that they did not neglect, and the immigration of the yellow labourers received a great impetus despite the fact that Russian prestige demanded a reduction in their numbers and importance. It was always easy to cross the frontier unperceived, and the Russian bank of the Amur and Ussuri was more popular than the opposite bank in their own territory. In 1904 the proportion of Chinese to the Russian population in the Amur and Ussuri districts was 16 per cent. In 1908 it was 24 per cent,

Unlike the Koreans they do not attempt to settle permanently on the land. In Transbaikal the spread of Chinese immigration was even more decided, especially in 1909. The Chinese merchant is found as far west as Irkutsk, and in the country east of Baikal the Chinese small trader reigns supreme. Everywhere much labour is wanted and the Chinese are indispensable. Nor will they be less indispensable because the Government orders that they shall be dispensed with. A passage from a Russian author will illustrate the way in which Russian needs are met by labour and commodities from over the Chinese frontier: 'A man in Khabarovsk, for instance, lives in a house built by Chinese labour of Manchurian timber: the stove is made of Chinese bricks. In the morning the Manchu *vanyka* brings water from the well. In the kitchen the Chinese boy gets the Tula *samovar* ready. The master of the house drinks his Chinese tea, with bread made of Manchurian flour from a Chinese bakery. The Chinese and Koreans come and offer their produce, eggs, vegetables, fruit from Shanghai and so on. The boy runs to the bazaar to fetch Mongolian meat and cooks the dinner. The mistress of the house wears a dress made by a Chinese tailor, and the master gets into his *chetchuncha* when the warm weather begins. In the yard a Korean is at work chopping wood.'

Koreans

The Koreans are an important element among the immigrants. They began to cross the border in 1860. Bad years of harvest and the extortions of the official class had driven them from their country. In 1869 there was wholesale immigration from Korea to the Ussuri district, and after that there has been a steadily increasing influx. In this country there is a great deal of undeveloped land, mostly prairie with scattered trees. This land is in the hands of Cossacks who live in the villages and spend their time in the taverns while the Koreans to whom the land is leased actively develop it. They are more efficient farmers than the Russians and their results are correspondingly better. They do not blend at all with the Russian inhabitants, for instead of living in villages they dwell in tents in the middle of their fields. Though pre-eminently agriculturists, some of them settle in the towns near the scattered gold-mines.

In 1882 there was an edict that none but Russian subjects should acquire land in Siberia : only in exceptional circumstances could the Governor-General give leave to foreigners. An agreement was made with the Korean Government that Koreans who had immigrated before 1884 could be admitted to Russian citizenship, and later arrivals could remain in the country for a short time, but must then sell their immovable property and return to Korea. This was not put in force till 1891, when many Korean settlers were given land by the Chinese in Manchuria. Besides the temporary occupation of land for short periods, there is an annual migration of Koreans who work in these provinces from the spring to the autumn, but return home for the winter.

Japanese

The immigration of the Japanese has been mainly after the Russo-Japanese War. The Japanese artisan is to be found throughout eastern Siberia ; in crafts he is superior to the average Siberian, and immeasurably so in diligence, sobriety, and general trustworthiness.

The Hunghuses

There is a further danger which menaces alike the Russian settlers and the orderly yellow inhabitants. This comes from the Hunghuses, bands of robbers, whose origin is uncertain. It seems probable that they have been mainly recruited from criminals escaping from justice and other dregs of the Chinese population, who were attracted to northern Manchuria by its remoteness, and, when there, took to gold-mining, a capital offence when unauthorized. At first a disorderly rabble, they are now armed with Mauser rifles of German military pattern. They have long been a terror to the peaceable Chinese occupants of the district between the Ussuri and the

Amur. They have penetrated into the Russian territory of the Maritime Province, drawn thither by the attraction of the poppy-beds, and the desolate character of the interior of the province has given them many opportunities of working mischief, though they stand in much more awe of Russian authority than of Chinese. However, with the withdrawal of the bulk of the male population, owing to the European War, and the weakening of the village fortifications, the Hunghuses were greatly strengthened. There have been sporadic raids on Russian settlements, culminating in a serious attack on the port at Olgi Bay, to repel which a Russian armed force had to be landed.

Russian Views on Yellow Labour

Russian writers show great alarm at the growth of yellow labour and its competition with that of the white man, who has to be fetched from a greater distance and whose standard of comfort and proportionate demand for wages is higher. It has been pointed out that Russian labour can be brought into competition with Chinese by the introduction of labourers from the west, who have no intention of settling. Some hundreds are brought by the special emigrant tariff; they arrive in the spring and leave in the autumn, and earn several hundreds of roubles more than the cost of their journey. For public works it is no good making a demand of labour from the colonists: they are far too busy during the first years in meeting their own requirements, and cannot spare time and labour even under the inducement of good pay. The Koreans are regarded as more dangerous than the Chinese, because of their desire of settling permanently on the land. The alarm takes the form of a demand for a huge staff of supervisors and inspectors, and a strict registration of the number of yellow immigrants. A partial attempt in 1906–7, in part of the Primorsk Province, showed that there were in one region only of the Ussuri district 14,000 Korean Russian subjects and 26,000 Korean foreigners, and these numbers should probably be increased by 10,000, as the enumeration only took place in the most populated southern portion of the district. Other demands have been that those suffering from contagious diseases shall be prohibited from entrance, that the frontier shall only be crossed at certain points and then with the sanction of the Russian consul, that annual permits shall be held and paid for and a fine exacted if they are not renewed. But the most insistent demand is that in all Government enterprises Russian labour shall be made to replace the labour of the yellow man, even though it cost more. The question of their settling on Crown lands is also urgent: many Chinese and Koreans are said to be settled owing to some illegality, and Russians do not view with equanimity the occupation of a large amount of agricultural land by yellow men.

Encouragement of Colonization in Arctic Russia

Since 1876 the Russian government has been encouraging the colonization of the Murman coast and has offered many inducements to settlers, whether Russian born or naturalized foreigners. Among other privileges these settlers were exempted from actual military service and passed into the naval reserve. They were excused the payment of state taxes and were allowed to receive, without duty, foreign goods imported in Russian or foreign ships direct to the coast. But this regulation had to be modified, for it led to the wholesale importation of intoxicating liquors and general drunkenness. In consequence the importation of foreign spirituous liquors was prohibited. Other privileges granted to Murman settlers were: state loans of from £5 to £15 granted at the discretion of the government and repayable in six years; free timber for building or a subsidy of £10 to £20 for the purchase of timber; and the right to hunt fur-bearing animals and to fish without licences. These conditions applied also to any nomadic Lapps who wished to settle definitely on the Murman coast. Finns and Norwegians were the earliest settlers to be attracted, and later came Russians. The establishment of a regular line of steamers between Vardö and Arkhangel, under government subsidy, helped the movement. The Finns and Norwegians kept to the west and the Russians farther east. There are Finn villages on Bolshaya-Volokovaya Bay on the west of Ribachi Peninsula. Many of the settlers are in villages on Pechenga Gulf, where the Pechenga Monastery is situated. Teriberskaya Bay and Gavrilova Bay are well peopled. Kola Inlet is another centre of villages, and here are the most important settlements on the coast, Kola, Murmansk, and Alexandrovsk.

Preponderating Advantages

A writer in the Journal of the Arkhangel Society (Mr. O. M. Latkin), in 1912, strongly advocates the colonization of the extreme north of European Russia. Against the rigorous climate and the absence of means of communication he sets the great natural wealth of the region: its mines, meadows, forests, and its rich supply of animals, birds, and fish.

Conditions of colonizing Crown Lands

In order to attract educated brains for the development of these resources he feels it essential to allow a free choice of land from the Crown provinces on the following conditions: (*a*) Crown land should be valued proportionately from 10–20 roubles (£1 1*s.* 4*d.* to £2 2*s.* 8*d.*) a desyatin (i.e. the value of the rent for 100 years—at present the Crown receives 10 or 20 kopeks (2½*d.* to 5*d.*) a desyatin for the northern forests); (*b*) unreserved rights of working the forests or mines should be granted, with the proviso that for forest material a tax should be levied according to the already existing tax, of which 30 per cent. should be paid by the owner to the Crown in part payment till the whole is paid; the remaining 70 per cent. should be applied for the benefit of the owner to develop such industries as tar-boiling, pitch-distilling, cattle-breeding, or agriculture.

The right of owning large tracts should be distributed among all classes; large tracts should be allotted at distances of not less than 40 or 50 versts (26 to 34 miles) from one another, so that new settlers or the aborigines in the districts between should have good models of reformed methods of agriculture to copy.

Roads are the first essential condition.

There is likely to be obstruction to the colonization of the Kanin and Timan tundras from the Samoyedes on the strength of an edict of Ivan Vasilovich (April 15, 1545), but there should not be a pedantic insistence on the terms of an obsolete edict in view of the good which will be done to the Samoyedes themselves, who, having lost nearly all their reindeer and being reduced to abject poverty, could learn something of rural economy from their new neighbours, and would fare as prosperously as the Samoyede village of Kolvinsk on the River Usa in the Pechora district.

The writer referred to sees no insurmountable obstacle to the colonizing of the whole of the north, from the coast of Norway to the mouth of the Yenisei, and suggests the employment of convict labour to effect his object.

Various proposals have been made for the construction of railways in north-eastern Russia, with a view to linking the Ob navigation with ports on the Barents Sea and facilitating Siberian trade. Though these railways would be mainly concerned with through traffic they would help to open up the country and overcome some of its greatest disadvantages to colonists.

CHAPTER VII

RELIGION IN SIBERIA

Russian Religion: History—The Clergy—Church Government—The Orthodox Religion—Raskolniki—Shamanism.

Russian Religion

History

The Russian Orthodox Church is a branch of the Eastern Church. In 988 Vladimir was converted and had his subjects baptized in platoons in the Dnieper. For a long time the Russian Church was in close dependence on Byzantium, but with the fall of the Eastern Empire Russia, who had gradually asserted her independence, took her place as defender of the faith: in the sixteenth century the patriarchate of Moscow (transferred from Vladimir and before that from Kiev) was recognized by the patriarch of Constantinople. Nikon, patriarch of Moscow, in the middle of the seventeenth century, carried out a series of reforms consisting largely of the correction of errors which had crept into the rites and liturgical books; but, although he insisted that he was only reverting to the practice of the primitive Church, his reforms were met with much vehement opposition led by the Tsarina. The *Starovyeri* (Old Believers) would not consent to such changes as the use of Alleluia three times (in honour of the Trinity) instead of twice (in recognition of the human and divine nature of Jesus Christ), nor for the same reasons to the use of three instead of two fingers in giving the blessing. Anathemas were as ineffective as appeals to reason, and the fact that they were anathematized in the year 1666 made the Schismatics think of the number of the Beast in the Book of Revelation and anticipate the reign of Antichrist. From this time dates the great schism, which led to exile and the formation of many dissenting bodies. Peter the Great asserted his authority by keeping the patriarchate vacant for twenty years, and then, in 1721, establishing in its place the Holy Synod, consisting of ecclesiastics nominated by himself, with the Procurator-General, a layman, as Imperial Representative. The Holy Synod now contains the metropolitans of Moscow, Petrograd, and Kiev, the Archbishop of Georgia, and other bishops sitting in rotation.

The Clergy

The clergy are divided into Black (regular or monastic) and White (secular), whose brown habit belies their name. The former must be celibate, and all bishops and high dignitaries of the Church are drawn from among them. Their ranks are *archierei* (metropolitans, archbishops, and bishops), *archimandrit* (abbots), and *igumen* (priors): below these are the monks. The white clergy are divided into protopopes or *protierei* (parish priests of the largest churches, or of churhes with others under them), popes or priests, and deacons, lectors, &c., the minor rank being recruited largely from intending students who could not pass their examination. The priests must be married; if a priest loses his wife he is not permitted to marry again. The parish priests are poorly paid and their lives are a constant struggle against poverty; but they have ample spare time, as they have practically no duties beyond the holding of services, and visiting their flock is entirely at their own discretion: so they can devote themselves to agriculture. In the remoter parts of the north-east the priests are known to act as commercial agents, and are sometimes at once the purveyors and the victims of drink. The village priest lives in a house built for him by the peasants and draws his scanty income from diocesan funds: there are no tithes, and the bulk of Church property was absorbed by Catherine the Great, though Nicholas I restored to the Church what had remained in the hands of the Crown.

Church Government

There are 66 dioceses (*yeparchia*) in Russia. The following are wholly in Asia: Tobolsk, Omsk, Tomsk, Krasnoyarsk, Irkutsk, Chita, Yakutsk, Blagovyeshchensk, Vladivostok, and Turkestan (of which the see is Vyerni); the following are partially so: Yekaterinburg and Orenburg. They correspond as a rule with the divisions into governments. They are divided into rural deaneries, which consist each of from 10 to 30 parishes, many of which, especially in Siberia, are very extensive and have scattered populations of several thousands. The parish church is usually under a *prikhod* (corporation) consisting of priest, deacon, two *diechoks* (bell-ringer and reader), and a widow who prepares the sacramental bread. A parochial council determines the sum due from each householder for the upkeep of the Church. The number of parishes in Siberia is being increased; arrangements were made for 101 new parishes to be supplied with priests in 1911.

Distinguishing Features of the Orthodox Religion

A few distinguishing points in the orthodox church may be noted. The Council of Nicaea in 787 is the last ecumenical council that they recognize; they do not admit the papal supremacy; they have the Scriptures in the vernacular, though it is not modern Russian, but old Slavonic, that is the language of religion and religious services, apart from the sermons, which are in Russian (the Slavonic language is taught in schools); they do not believe in the procession of the Holy Ghost from the Son as well as the Father, and omit the *filioque* clause from the Nicene Creed.

Most Russian churches are rectangular buildings with five domes, the largest being in the middle. The principal entrance is at the west, where there is usually a detached campanile. The church is divided by an ambo (or *ikonostas*) into nave and sanctuary. In the former stand the worshippers, there being no seats or benches. On the *ikonostas* are sacred pictures or *ikons*, in front of which lamps are burning. Pictures are allowed, and even bas-relief, but statues are forbidden. Opposite the central door through the *ikonostas*, which can be used by the priest alone, is the altar (*prestol*), on which is laid a New Testament and the host. It is forbidden to pass in front of the altar. During the services no instrumental music is allowed, but the beautiful hymns of the Eastern Church are usually sung in three parts by men and boys. The congregation constantly join in the appeal *Gospodi Pomilui* (Lord have mercy upon us). The communion is administered in both kinds to the laity, but the bread and wine are mixed together in a spoon. Children are given water and wine alone till they are seven and can go to confession. Mass is only celebrated once a day. Attendance at least once a year is enforced by law. Baptism is by immersion and anointing with chrism immediately after takes the place of confirmation. There are many fasts besides the regular fasts of Wednesdays and Fridays throughout the year. The chief ones are (1) Lent, (2) St. Peter's fast, from Whit-Monday to June 29, (3) the fast of the Virgin Mary, from August 1 to August 15, (4) St. Philip's fast, from November 15 to December 26. Besides this the monastic clergy always fast from meat. Festivals have a demoralizing effect ; much of the vodka-drinking is especially connected with them ; till 1907 it was a crime punishable by law to work on a holiday.

Raskolniki (*Dissenters*)

The differences between Popovists and Bezpopovists. have been described under the account of colonization (see p. 190). The former made advances towards reconciliation in 1862, on the basis of accepting orthodox priests, but retaining the unrevised books. This led to further division: There were now (1) those who recognized the metropolitan and this compromise, (2) those who recognized the first but not the second, (3) those who recognized neither. There are some offshoots of the Bezpopovists : (1) the *Philippovsti*, followers of one Philip who burnt himself in 1743 : these exalt self-immolation into a principle ; (2) the *Stranniki* (pilgrims) ; (3) the *Byeguni* (runners): both these reject legal marriage; (4) the *Nyetovsti* (deniers) deny the necessity for common worship; (5) the *Molchalyniki* (mutes) will not utter a syllable under torture. Besides these there are (1) the *Khlusti* (flagellants), who subdue the flesh, but indulge in ecstatic forms of worship: they are a secret society and are nominally members of the Orthodox Church. They were founded in 1645. (2) The *Doukhobors*, 'spiritual fighters,' who hold conscientious objections against military service, and were therefore made to live in Trans-Caucasia. When service became compulsory there they went to a home found for them by the Society of Friends in Canada, where they have lived honest and industrious lives. (3) The *Molokani* (milk-drinkers), founded in 1765. The name is given them because they drink milk during fasts, but their tenets resemble those of the Quakers. (4) The *Skoptsi* (eunuchs), who advocate castration for the 'kingdom of Heaven's sake'. The more moderate of them allow absolute chastity to take the place of self-mutilation. These last two sects are the most prominent in Siberia, the former being found along the Amur, the latter in the Yakutsk Province.

Other forms of belief

Of other creeds there are in all Russia about 12,000,000 Roman Catholics and 7,000,000 Protestants (including Finland), 5,000,000 Jews, 14,000,000 Mohammedans, and 500,000 Buddhists. There are Lutheran and Roman Catholic churches in Omsk, Tomsk, Krasnoyarsk, and Irkutsk. Jews are found in great numbers at Kainsk, Buddhists among the Buryats in Transbaikal, and Mohammedans not only among the Tartar tribes, but also to a considerable extent among the Cossacks in the neighbourhood of Petropavlovsk.

Shamanism

From end to end of Siberia, despite the ban of the Russian authorities, the primitive religion of northern Asia survives and exercises the strongest influence upon the inhabitants. Converts to Christianity and Buddhism still resort to it, and it even affects Russian officials and peasants who come to the country, and has among its adherents many of the half-breeds who have grown up in the land. It has the character of a primitive religion, and is probably much older than Buddhism. It may have been the earlier religion of the Mongolian people, south of Siberia. It has more magic than theology about it, and its doctrinal core is not very great; it is in its forms and outward manifestations that it is really important. It is always difficult to acquire exact information from the Shamans themselves, who are in fear of the Russian authorities; and the ritual of Shamanism attaches itself readily to the different racial religions and mythologies.

Religious doctrines of Shamanism.—It is therefore extremely hazardous to venture on a general description of the religious beliefs that underlie Shamanistic forms, but perhaps it may be summarized as follows. The universe consists of a number of layers or strata, separated by a kind of intermediate space or matter; there are seven upper layers, which constitute the kingdom of light, while below the earth are seven or more nether layers, which form the kingdom of darkness. Between these lies the earth, subject to influences from above and below: above are the good spirits (*aiy*), below are the evil spirits (*abassy*). In the seventh layer above, in high heaven, reigns Ai Toion, perfect and good; in the fifth (or the ninth) below is Erlik Khan, the Ahriman of this dualistic system. Shamans alone possess power over this spiritual world, and this power is exercised more over the bad than the good, not necessarily for evil ends, but because the good spirits need less propitiation and are naturally inclined not to do mischief. In accordance with the nature of the spirits over whom they exercise influence Shamans are known respectively as white and black. In fact the whole faith has been known as 'black faith' in opposition to 'yellow faith', i.e. Buddhism. These spirits are largely the spirits of ancestors, and so Shamanism is closely associated with what constitutes so large a proportion of primitive cults, viz., ancestor-worship. Shamans among the Yakuts (the tribe among whom this cult is most fully developed) have three spirits, *ämägyat*, which is indispensable (the same name is applied to the iron breast-circle of the Shaman, which he wears as a symbol of his office), *yekyua*, which is hidden away but incarnates itself at times in animals, and *kaliany*, impish and mischievous, a sort of secondary personality.

Derivation of name.—The word shaman is uncertain in derivation. There is a Sanskrit word *śramaṇa* (= religious mendicant), the Pāli form of which, *sramana*, has the same meaning. On the other hand there is a Manchu word *saman* which means 'one who is excited' and this word is found among the Tungus.

Differences between Palaeo-Siberian and Neo-Siberian Shamanism.—There is a considerable difference between Shamanism as practised among the Palaeo-Siberian tribes (Chukchee, Koryaks, &c.) and among the Neo-Siberians (Yakuts, Tungus, &c.), though the influence of the Tungus, widely spread among other tribes, has greatly modified the Shamanism of north-eastern Siberia. With the Palaeo-Siberians 'family' shamanism is more frequent than 'professional'. The head of the house will shamanize, and in the absence of the father (e. g. among the Chukchee) the mother will take his place as family shaman. There is also found a kind of communal shamanism. Among them also women are more important than men as shamans; women being of a more nervous and excitable temperament are more natural recipients of the shamanistic gifts; it may be noted that women are the most frequent victims of Arctic hysteria (*menerik*). On the other hand the Yakuts assign an inferior position to women, and will only resort to them as shamans in the absence of men. Further, among certain Palaeo-Siberians (Koryak, Kamchadal, Chukchee, and Asiatic Eskimo) appears that extraordinary phenomenon, the supposed change of sex, by which men come to behave as women and women as men. The change goes through various stages; the person who undergoes the transformation will first manifest the change by arranging his hair as a female; then he will adopt woman's dress; then he will change his voice and his general habits, performing the occupations of a woman in the house; finally he will seek a 'husband' and live with him in homo-sexual relations, while at the same time often having a female concubine and begetting children. Public opinion disapproves of this homo-sexuality, but is discreet in its utterances on the subject, as such transformed shamans are held to be particularly dangerous. The 'soft-man' has his *kele* (spirit) husband, who will protect him: in fact even without sex-transformation a shaman often has a *kele* wife, as well as his own. The 'husband' chosen by the shaman is usually selected from among his near relations. Women are transformed to men in the same way, with the corresponding change of dress, voice, and occupation, and the adoption of a 'wife'. The Koekchuks of the Kamchadals have been especially described in their place. Another distinction between the Palaeo- and the Neo-Siberians is that among the latter the dualism of good and evil spirits with their attendant black and white shamans is much more emphasized. The 'white' shamans take part in spring festivals, marriage ceremonies, fertilization rites, and the curing of diseases among the Yakuts: the 'black' shamans deal with evil spirits, but are not necessarily malefic: they shamanize to assert their prestige; they foretell the future; they call up spirits and wander into spirit-land. The Buryats speak of great 'contests' between black and white shamans, who 'hurl axes at one another at a distance of hundreds of miles'. The black shaman is not a popular person, and is sometimes killed by the inhabitants. The grave of a black shaman among them is guarded by an aspen, and the body is fastened to the earth by a stake taken from this tree. Among the Samoyedes there is no distinction between black and white shamans. It will be seen later that there are marked differences in the actual performances of the Palaeo- and Neo-Siberians.

The 'call'.—Shamanism is sometimes hereditary. This is what would be expected both from the recurrence of mediumistic gifts in families and from the greater ease of keeping secrets known to members of the family alone. A shaman is often subject to hysteria, but can control himself between the fits. Before entering on his vocation a shaman has often had severe nervous affections. The 'call' comes in various ways. We hear of a Gilyak boy of twelve, who had his call during a deep sleep, and when awakened threw himself about and spoke with different intonations. To the Tungus a former shaman appears in a dream. An Ostyak will sell to another the spiritual gifts that have come to him. Among tribes in the Altai the call is involuntary. Among the Chukchee there is great fear of the 'call': the Chukchee youth is afraid that it will come to him; when it has come, he segregates himself and is abstracted in manner, he sleeps much, and is carefully guarded lest he should freeze to death during sleep. The Buryat child is supposed to be called at a very tender age: its soul then, it is imagined, goes away to be trained among the 'West Tengeris' if he is to be a white shaman, among the 'East Tengeris' if he is to be a black shaman. When he reaches adolescence, certain symptoms begin to be revealed. Among the Samoyedes the novice, at about the age of 15, is entrusted to an old shaman.

Novitiate and training.—A 'call' to a shaman means that he has come under the protection of one or more spirits: his eyes have a distinctive appearance: the expression is said to be a combination of shyness and cunning, and it is alleged that the shaman can often be picked out from other men because he has this look. Long periods of preparation follow; the training includes lessons in singing, dancing, drum-beating, ventriloquism, and other tricks, and the power of concealing fatigue; and stages of consecration, which differ among the various tribes. The novice is conscious of the solemnity of his profession, and usually has a strong feeling that he has to consecrate his gifts for the good of his fellow men: he is told not to demand high prices from either rich or poor, and, if he is asked to attend to a rich and a poor man, to attend to the latter first. Considerable danger is felt to be attached to the profession, as 'the spirits will kill a shaman who in any way disobeys them' ; but there are compensations, for he is usually safer than anybody else from the anger of his fellow men, on account of the sacrosanct character of his calling. But no persecution will make him give up his shamanism : a shaman, whom Stadling met, who was a Christian nominally, used to confess once a year to the priest (and present him with a blue fox skin). It is worthy of note that the tribe which has probably developed shamanism more than any other is the Yakuts, who have been nominally Christians for upwards of 200 years.

Classes of Shaman.—Shamans are of various kinds. Among the Chukchee there are three kinds of professional shamans : the first practise ecstatic ravings, the second foretell the future as prophets, the third utter incantations ; these last again are subdivided into good and bad, and are distinguished by their red and black coats respectively. With the Yakuts they are divided into Great, Middle and Little Shamans in accordance with the degree of their powers ; the first has his *ämägyat*, from Ai Toion himself ; the second has *ämägyat*, but it is not of so powerful a kind ; the third has nothing that deserves the name of *ämägyat*, but is only an abnormal neurotic person, 'who can cure trifling illnesses, interpret dreams, or frighten small devils away'. Originally among the Yakuts there was more of woman-shamanism, as among the Palaeo-Siberian tribes. The Altaians, besides the shaman (*kam*), have other personages of a similar kind. These are (1) *rynchi*, who foretell the future during attacks of pain ; (2) *telegochi* (guessers) ; (3) *yarinchi*, who divine by the use of a bladebone ; (4) *kollkarechi*, palmists, who divine from the hand, (5) *yadachi* who control the weather by means of a stone (*yadatash*) found in a defile, where winds blow continually : to obtain this stone they must swear away all that they have.

Professional dress and equipment.—The shaman has a professional dress when engaged on his occupation. Among the Neo-Siberians the four most general features are the coat, the mask, the cap, and the iron plate about the breast. The costume is less complicated among the Palaeo-Siberians.

The Chukchee for instance have no special dress: they merely desire originality, and will wear any coat that they think will impress. Sometimes the Chukchee have adopted Tungus designs on the coat without knowing their meaning. The coat is most elaborate among the Neo-Siberians. Attached to it are pieces of metal each with a name and meaning of its own. Among the Altaians not all shamans are entitled to wear the coat and the cap. The mask is of skin, wood, and metal, painted and ornamented with a great beard. The iron plate (*ämägyat*) is handed down from shaman to shaman. The pieces of metal are supposed to have a soul and to be capable of resisting rust. Among the Yakuts there are hereditary blacksmiths who are associated with the shamans, and manufacture their properties. The most characteristic emblem of shamanism is the drum. A special meaning is attached to it by the Yukaghir: they call it *yalkil* (gulf), as the gulf into which the shaman dives to reach the spirits. So too the Eskimo think that the souls of the shamans descend into the lower world of the goddess Sedna. Some of the north-eastern tribes (e.g. Koryaks) strike the drum from below. The word for 'drum' is everywhere the same (*tungur*), whereas the coat has various names, which suggests that it is later. The drum is rare among the Buryats, who have one special accessory, viz., the horse-staves: of these two are of wood, two of iron, but the latter are only bestowed on a shaman after his fifth consecration. The wooden ones are cut for the novice. These horse-staves represent the horses on which the shaman takes his flight to the upper and lower worlds. There is also a *shire*, a box which contains the sacred emblems (horse-staves, &c.), which the shaman acquires the right of holding after his fifth consecration. The sun, moon, and secondary deities are represented on it. The Buryats also have a musical instrument used only by shamans, a sort of jew's harp called *homus*.

Shamanistic rites.—The nature of the shamanistic performances can be best realized by describing two: the first is as it would be given among the Chukchee, the second as among the Yakuts. The first is given when it is almost dark : the shaman begins to beat the drum softly and to sing plaintively : the song imitates the cries of animals, which seem to proceed from various corners of the house. Then suddenly the song ceases : when it is over, the shaman is found lying exhausted. Sometimes the shaman uses a hieratic language, a mixture of Yakut, Yukaghir, and Koryak words : shamans cannot remember what they recite in their semi-hypnotic state, and genuinely do not understand the language they use. A variety of conjuring tricks are performed.

Among the Yakuts the performance appeals to rather higher emotions. The shaman kneels on a white mare's skin, bows to the four corners of the earth, and sprinkles the ground with water from his mouth. After other rites, the shaman begins to play his drum, and utters wild cries imitating animals. Then he chants an incantation, the spirits come, the shaman falls, leaps, and dances ; members of his 'congregation ' hold him by leather thongs, lest the spirits should make away with him. In the south Yakut district he is unfettered. After these movements he approaches his patient, drives away the cause of the illness, and prescribes what sacrifices must be made to the powerful spirits whose servant he has banished. His prophetic gifts do not leave him at once, but he foretells future events. He goes a mystic and symbolic journey through the strata of the universe announcing the various points in his travel that he has reached.

Despite the trickery there is evidence of considerable mediumistic powers ; there is use of auto-suggestion in the trances and probably a skilful handling of hypnosis.

One of the earliest accounts of shamanism was published in China in 1747, written in the Manchu language. For the subject generally, see M. A. Czaplicka, *Aboriginal Siberia*, from which the above examples are taken.

CHAPTER VIII

HYGIENE

Climatic complaints—Zymotic diseases—Nervous diseases—Want of sanitation

Climatic Complaints

There is no reason for regarding the climate of Siberia, despite its rigours, as unhealthy, unless the nervous complaints mentioned later have any connexion with the winter darkness and cold. Settlers in Siberia, including exiles, suffer no ill effects from the climate, whether in the far north or in the agricultural regions of the south. The great cold of winter, it is true, is liable to cause frost-bite, but with adequate shelter and sufficient nourishment, the risk of this proving serious is small since strong winds are rare in winter, and in their absence great cold is quite endurable with safety. Snow-blindness is not uncommon in spring and autumn, but the lack of winter snow in many parts, the darkness of the northern winter, and the general prevalence of green trees to relieve the eyes minimize the occurrence of this trouble.

Zymotic Diseases

Apart from climatic influences, however, there is much disease among most native tribes in Siberia. Measles is common, especially among the Koryaks and Yukaghir, and has devastated whole villages. Smallpox is endemic in many parts: it is said to have caused a steady reduction in the numbers of the Yukaghir and Tungus. Tuberculous disease is prevalent, though seldom diagnosed on account of the absence of medical officers in most parts: it was very probably introduced by Russians. Sibe ian boil plague, a form of anthrax which also attacks cattle, is found from the Urals to the Chinese frontier, especially in summer. It occurs in both external and internal forms. The latter is generally fatal in one to four days. Goitre is reported principally from the Lena valley, where it is most prevalent among women, and from the Amur region. Syphilis is rampant throughout Siberia. Ophthalmia and other diseases of the eye are very common: some are due to snow-blindness, as noted above, others to venereal disease, and others to the smoky nature of the interior of all native huts. Leprosy occurs in the Lena, Kolima, and Amur regions, Sakhalin, and elsewhere. It is generally associated by the Gilyaks with eating fish, especially one species of salmon. Cholera is never absent from the Amur and Maritime Provinces, and every few years assumes the proportions of an epidemic. In 1910 a violent epidemic of cholera raged in Vladivostok, Khabarovsk, Blagovyeschchensk, and Nikolaevsk: most of the victims were Chinese and Koreans. In the same year an epidemic of plague introduced from Manchuria into the Amur region was successfully fought by the use of injections. The last epidemic of typhoid in the same region was in 1908. Thus the Amur and Maritime Provinces seem to be the least healthy parts of Siberia owing to their proximity to any diseases rampant in China, Korea, and Japan.

Nervous Diseases

Certain peculiar forms of nervous affection are common among the natives of Siberia, and are known collectively under the name of Arctic hysteria, because, as far as Siberia is concerned, they are confined to the polar and subpolar regions. They are, however, closely akin to nervous affections of the natives of Java, Abyssinia, South Africa, Madagascar, Brazil, Peru, and elsewhere, and particularly the Malays, and so, properly speaking, are not peculiar to Arctic lands, although they are probably accentuated by the darkness, and are certainly most prevalent during winter. Hardships increase the occurrence of hysteria: in times of famine whole villages may suffer from it. It is noticeable also that sedentary people suffer far more than nomadic people, who are better inured to hardships.

Arctic hysteria is most prevalent among Neo-Siberians, but is found also in certain forms among Palaeo-Siberians. European exiles from Russia to Arctic Siberia seldom suffer from it.

Many forms of nervous disease are included under this head, quite apart from the hysterical manifestations of the shamans, which are looked upon by the natives as an inspiration, but are not impossibly connected in origin with Arctic hysteria (see Chapter VII). A common form is expressed in timidity and fright, with an inclination to repeat all visual and auditory impressions. Another type is brought on by sudden shock or pain, but it is sometimes periodic and recurs without apparent direct cause. The patient is afflicted with spasms or falls into a trance, howls or dances, and the fits are often followed by extreme exhaustion or prolonged sleep for several days. In some symptoms this form closely resembles epilepsy. It is sometimes accompanied by manifestations of erotic mania, and by Yakuts, Yukaghir, and others is ascribed to the influence of evil spirits, but looked upon as a disease. All these forms seem to be commonest among people who have recently moved into Arctic regions. Melancholia and so-called voluntary death are other forms of hysteria. Melancholia occurs chiefly among people domiciled in Arctic Siberia, and is not uncommon among the inhabitants of other far northern lands. Voluntary death is also common among the tribes of northern Siberia. Old or infirm people request their relatives to put them to death: this, however, may be the outcome of a desire to escape suffering, and it is doubtful if it can be looked on as a form of hysteria.

Want of Sanitation

The difficulties of combating disease throughout Siberia are very great. Most of the natives are dirty and devoid of the most primitive ideas on sanitation. Owing to the vastness of the land over which they roam, no adequate medical supervision is possible. At the same time it must be remembered that the state of health of the majority of the tribes in Siberia has little influence on the Russian settlers who, by the nature of the land, are more or less confined to certain regions, where measures of preventive medicine, even if difficult, are not impossible. The state of the Siberian towns (see p. 313) leaves much room for improvement in matters of sanitation and health.

CHAPTER IX

AGRICULTURE

Western Siberia—Eastern Siberia

Western Siberia

The peasant population of western Siberia constitutes about ninety per cent. of the whole, and it is nearly all engaged, directly or indirectly, in agriculture. The agricultural districts of western Siberia are the Government of Tobolsk, excluding the districts of Berezov and Surgut, and the Government of Tomsk, excluding the district of Narim, but in the steppes agriculture is only practised successfully in parts of the districts of Kokchetav, Atbasar, and Petropavlovsk in the Akmolinsk territory, and in the Semipalatinsk, Pavlodar and Zaisan districts of the Semipalatinsk territory. In 1911 over 10,800,000 acres were sown in Tobolsk and Tomsk, i. e. 36 per cent. of the whole sown area of Asiatic Russia. In the steppes over 5,400,000 acres were sown.

Cereals, root crops, and fodder

Three zones of agricultural land.—(1) North of lat. 58° N. there is a region in which agriculture is only sporadic. It consists largely of *urmans* (swamps), which are quite unsuited to tillage ; the arable lands are either the more elevated parts of the river valleys, which are not submerged, the *uvals* (inclined banks), or level ground surrounded by *yars* (abrupt precipices). North of lat. 60° N. hardly any cereals are grown; corn is planted as far north as Berezov (about lat. 64° N.), but is of no economic importance in those parts.

(2) Between lat. 56° N. and lat. 58° N. the region is an almost perfectly flat plain, with deciduous trees predominating

over conifers. Swamps, though extensive, occupy less of the total area: there are plateaux as well as river valleys used for agriculture. Lands suitable for cereals are often spread over great tracts. The river valleys are hardly ploughed at all (except for the *uvals*), as being too liable to inundation.

(3) South of lat. 56° N. the region is variegated with many small lakes divided by ridges with sloping sides. There are islands of fertility with flat spaces between that are entirely barren. The wild cherry growing on unploughed soil is a sign of its fertility. It grows on the west Ishim steppe, but not on the east Ishim and Baraba steppes. Much of this region is particularly adapted for wheat. The west Ishim steppe is the most fertile region; the Baraba and east Ishim steppes closely resemble one another. The Baraba steppe has great variations in itself; it is least fertile in the north, where it becomes swamp, most fertile in the south, where it reaches to the foothills of the Altai. In the Kirghiz steppes few cereals are grown, as the Kirghiz eat little bread of any sort, but a great deal of meat.

Soils.—In the river valleys there prevail very sticky clayey soils, partly grey, slightly tinged with humus, partly black. There are two kinds of black soil: (1) argillaceous *chernozem* or black soil, on the raised ground, the most fertile of all the soils, (2) a black earth which is poor and barren and of a peaty character, only adapted for oats. In the middle region especially there are what is called *byeliks*, where a very thin layer of turf (2 or 3 inches thick) lies over a stratum of almost unproductive light-grey clayey soil (9 or 10 inches thick), superimposed on a reddish-yellow clay. These are only of use with manure, and have to lie fallow for twenty-five years after three or four crops, as only the top layer nourishes cereals. In the south region the predominating soil is a dark brown, friable, clayey soil, with a large admixture of white sand upon a reddish clay subsoil. In the Baraba steppe upon the broad sloping ridges black soil is everywhere, and on the narrow and more abrupt ridges a clayey soil. The soil along the railway has a loamy substratum with a surface of black earth (14 to 24 ins.) which is entirely stoneless.

Crops vary in accordance with varieties of soil. Wheat grows best in south-west Tobolsk, the Kainsk and Mariinsk districts of Tomsk, the Altai and parts of Akmolinsk territory where there is a sandy black soil; in all these parts it is more than half the grain sown; barley and spring rye are adapted to the brownish soils of the central part of Tomsk, where the soil is infertile; east from Tomsk grows winter rye, and oats are commonest along the great Siberian road. In the northern districts winter corn occupies the chief place. In all Asiatic Russia spring corn considerably predominates over winter corn.

Methods of cultivation.—In the agricultural parts of Siberia it is usual to sow the land for two or three years, and then for a year to leave it fallow; after this to sow it again for one or two years, and to repeat the process till it shows that it needs a rest. Then, when certain signs known to the peasants present themselves, it is reploughed. During the early part of the period the more exhausting grains are sown, such as wheat and rye; towards the end of the period, barley and oats. In the region of Lake Zaisan there is intensive culture by irrigation There are irrigating canals (*aryks*) from which little runlets are taken all over the fields. The water is let on first before ploughing, and then, according to the weather, from two to four times more, while the plant is growing. After eight crops the field requires either three years' rest or manuring. It is always the same crop, because during the harvesting the old seed would drop into the soil and spoil next year's crop.

Agricultural improvements.—Many improvements have been made in the system of agriculture in western Siberia. Fertilizers are employed to a greater extent, and there are many mechanical appliances introduced. A large amount of agricultural machinery is imported to meet a growing demand. Large wholesale purchases are made, especially of reapers, mowers, and rakes. The United States, with a widespread organization in the villages, have an almost complete monopoly, except for ploughs, which have hitherto been usually of German make. Chelyabinsk, Novo-Nikolaevsk, and Omsk are the chief centres of distribution, and there are others at Kurgan, Petropavlovsk, Pavlodar, Semipalatinsk, Tatarskaya, Karachi, Kamen, Barnaul, and Biisk. At a station on the River Irtish, near Omsk and also at Novo-Nikolaevsk, experimental work is being carried on ; arrangements are made under the Colonization Department for testing kinds of machinery and implements introduced into Siberia. Specimens of suitable machines are on view for farmers who visit the place. The value of sales was in 1910 £497,000 ; in 1911 £680,000 ; in 1912 £840,000 ; in 1913 £730,000. The purchaser usually makes the first half of his payment during the first year and the rest in instalments. Bad debts are rare. There are many agricultural depôts scattered through the country, organized by the Government ; but the private store has some advantages in the peasant's eyes over the Government depôt ; he has it more at his mercy, for it gives him long credit, and he can keep it waiting for his money, whereas he is at the mercy of the Government depôt himself. All agricultural implements enter the country free of tax.

Grain elevators are being erected at the cost of the Treasury, but in order to secure them more speedily the Siberian cooperative societies are undertaking the construction of elevators at their own expense. At Aleksyeevsk in Omsk district an elevator of 1,600 tons capacity has been completed ; it is provided with all the necessary grain-drying and cleaning appliances. This is the case, too, with the station at Mishkina. Eight credit societies are erecting an elevator of 6,025 tons capacity at the station of Kochenevo. At Omsk and at Kulachinskoe in the Omsk district grain stores are to be built. A large elevator of 16,070 tons capacity is projected for Novo-Nikolaevsk.

Agriculture has been stimulated along the line of the great Siberian road owing to the fact that this was formerly the one artery of commerce, and a large population was wanted for the work of transport and innkeeping. But the road has formidable competitors in the steamers upon the rivers and in the Siberian Railway; its monopoly has gone, and a large proportion of the population has taken to agriculture.

Milling.—Most of the cereal products are exported from Russia in the form of grain, and so milling has suffered. But the Tomsk Government contains a fair proportion of mills, among which wheat mills predominate with a return of £600,000. Flour mills are found principally in the districts of Tomsk, Biisk, and Barnaul, and in the neighbourhood of Novo-Nikolaevsk. The mills are well set up.

Export of cereals.—Siberia supplies the deficiencies of European Russia in bad seasons, so as to enable the Russians to export the reserves which would otherwise be retained for home consumption. Much goes to the non-agricultural parts of Siberia, especially the mines. Some was formerly used for vodka and other spirits. But the export of Siberian cereals is not great in proportion to the output of the country, and in comparison with that of European Russia it seems negligible. The cost of transport is very heavy. A German authority states that the cost of transport of a ton of wheat from Chelyabinsk to the mouth of the Rhine amounts to £2 16*s.*, while the carriage of the same quantity from India costs only 12*s.* There is still a large amount of land of great agricultural value unsown, and there are considerable openings for trade, especially if a serious effort is made to replace the German trade in ploughs.

Hay.—In 1914 in western Siberia 12,892,918 acres were under hay, with an output of 7,358,100 tons; in the steppes there were 9,271,773 acres, with an output of 3,546,500 tons. Hay-cutting is done near the big towns and along the post roads.

Potatoes.—Potatoes are grown further north than the limit within which cereals are profitable, e. g. at Samarovskoe, at the confluence of the Ob and Irtish. Other crops grown in the same region are rape and cucumber.

Fruit is of little importance. Some cherries are grown at

Kurgan ; in Tyumen there are a few small fruit gardens, and apples are grown at Tomsk, under the care of the University. Melons and water-melons are grown by the peasants for their own consumption.

Flax and hemp.—Flax is grown almost universally ; hemp especially in the black earth districts. These are grown mainly for the seed, whence comes an edible oil, which is of great importance for Russians as a substitute for butter on fast days. The fibres serve especially local needs for ropes and linen. The flax is good, but weak : the peasants cannot handle it properly, and it has but small international value. In the Barnaul district the hemp is of low quality and badly worked up. It is estimated that this district might produce 4,800 to 8,000 tons of hempseed and the same amount of linseed. At Omsk there is a steam oil-seed and colour mill, which manufactures various oils from linseed, hemp, and sunflower seeds, to the extent (in 1901) of 645 tons, and colours to the extent of 81 tons.

Sunflower.—From the seeds an oil is extracted and used for the same purposes as those from hempseed and linseed. It is grown especially in the Altai and at places along the Irtish.

Tobacco.—Altogether in Siberia in 1914 there were 18,198 plantations of tobacco, with an area of 1,984 acres. They yielded 4 tons of Turkish tobacco and 1,570 tons of lower quality. The chief tobacco-growing districts of Siberia are along the Irtish, south of Omsk, in the Cossack settlements, where about 160 tons are produced a year. The plantations are all small, and most of the work is done by women. About 30 per cent. is consumed locally ; the rest goes to the neighbouring fairs and to Omsk, where there is a tobacco factory, doing very good business, chiefly in cigarettes (£40,000 in 1903). The common Russian tobacco is *mahorka,* which is grown in Omsk and Petropavlovsk districts, and is chiefly used in the manufacture of cheap cigars. The peasant is showing a tendency to abandon *mahorka* in favour of cheap cigarettes. The import of tobacco goods by rail is increasing.

Woodland produce

Cedar cones.—The *Pinus cembra* produces a cone which is much esteemed for the oil extracted from its seeds. It is found especially in the Narim district, the northern parts of Tomsk and Mariinsk districts, and the mountainous parts of Biisk and Kuznetsk districts. The chief market is Tomsk. In a good year, which is once in four or five, 4,800 to 6,400 tons are gathered, the season being from the middle of August to the middle of September. The price is from 3*s.* to 5*s.* per pud ; the average turnover of a labourer is from about 12*s.* to £1 18*s.* In order to obtain the cone the trees are cut down if they are difficult to climb.

Wild fruit.—Great masses of bilberries and cranberries are exported from Turinsk, Tara, and Tobolsk districts. Rock-cherry, obtained in Tomsk district, is dried and ground to flour, and on fast days is boiled with honey and water and eaten as a kind of jam. The nut trade is especially developed in the Surgut, Tara, and Tobolsk districts of Tobolsk and the Narim and Mariinsk districts of Tomsk. In the Kuznetsk and Biisk districts the nomads do the nutting.

Mushrooms are put on the market, dried and salted.

Eastern Siberia

Cereals, root crops, and fodder

In spite of the climate agriculture is making rapid progress in eastern Siberia, especially since the Government began to encourage the transfer of the land from the community to the peasants. The settlers generally seem to make a decent living and to improve in appearance. When the first difficulties have been overcome, they are better off than they were in Russia. The land which is most accessible and most favourably situated for agriculture has already been occupied. New immigrants can no longer hope for natural grass-land, but once the taiga has been cleared, the soil is generally good.

Manure is only beginning to be used in some parts of

Transbaikal. The ramparts of it that often stand round Siberian towns and villages are left undisturbed. When the old land is exhausted, the peasant ploughs up a new plot. But now that he usually owns his land, he may be willing to make improvements, and adopt a less wasteful system of agriculture. A system of rotation of crops is generally followed ; it varies in different districts.

Eastern Siberia is by no means self-supporting in agricultural produce. Some 200,000 tons of wheat enter the Transbaikal, Amur, and Primorsk Provinces annually from Manchuria alone, while Irkutsk, Yeniseisk, and Yakutsk receive large supplies from western Siberia.

The chief peculiarity of agriculture in these provinces is the great preponderance of spring over winter grain. Even in the Transbaikal, Amur, and Maritime Provinces spring grain is less than 1 per cent. of the total. The long winter and slight snowfall make it almost impossible for winter grain to ripen.

Yeniseisk and Irkutsk.—Though cattle-raising is the chief occupation, agriculture is steadily extending. Spring crops predominate, but winter corn is sown on 28 per cent. of the cultivated land. In Irkutsk especially rye is the chief crop, then oats and wheat. Buckwheat, millet, potatoes, lentils, flax, and hemp are grown. In Achinsk rye and winter wheat do well because of the early winter and deep snow. In the Minusinsk district there is an area of dry steppe, surrounded by rich black earth, where a considerable agricultural population prospers. The warm dry summer enables spring wheat to ripen here before the autumn frosts. Beetroots and watermelons do well near Minusinsk. Hay is good and plentiful throughout these provinces.

The Minusinsk region alone has a trade in flour with Krasnoyarsk and the lower Yenisei. Some 4,000 tons are produced by about six steam or water mills, none of which has a capacity of more than 1,600 tons. There is a small steam-mill at Krasnoyarsk, one of 1,600 tons capacity at Cheremkhovskoe, and a couple of good-sized mills in Irkutsk and the neighbourhood. There are also some 20 water-mills for local use along the Yenisei.

The soil round Minusinsk is particularly favourable to beets, and a beet-sugar manufactory has been established there.

The Provinces east of Lake Baikal.—The territory east of Lake Baikal still needs large quantities of imported corn. Some 200,000 tons of wheat enter the Maritime, Amur, and Transbaikal Provinces every year from Manchuria for the use of the people and the troops. The Government is always the largest buyer. It is estimated that the Priamur alone could produce at least 600,000 tons more grain than it does at present. The Government proposes to put a duty on Manchurian corn, in order to encourage local agriculture, and to lower the freights on grain and flour from western Siberia. But it is held in Blagovyeshchensk, which is the agricultural capital of the Priamur, and the third in importance of the flour-milling towns of the Russian Empire, that, so far from excluding Manchurian wheat, this will only enable it to be sold at a higher price across the river. Indeed, the Russian mill-owners in Kharbin, who produce most of this imported flour, are in no way perturbed by the proposal. The campaign against Chinese labour has seriously raised the price of living, and the suggested duty will raise it still further.

The Colonization Department also intends to build grain-elevators along the Amur railway. The Government has been advised to build elevators of a capacity of 9,000 tons and a flour-mill of 18 tons capacity per diem at Aleksyeevsk. Smaller elevators are to be established at Bochkarevo, Bureya, Malinovka, Tigda, Yekaterinoslavka, and Gondatti. The extension of credit among the peasants is also to be encouraged. In spite of these suggestions and the advantages offered by the opening of the railway and other improvements in the means of communication, there is little prospect of these provinces being able to produce their own food for a long time.

Moreover, there is always the danger of over-production, with a consequent sharp fall in prices. Russia is one of the world's granaries, and it might not be easy to find a profitable market for surplus Siberian grain, owing to the inaccessibility of the country. Hence many authorities are of opinion that the future of the east Siberian peasant lies in stock-raising and dairy-farming rather than in corn-growing.

But the Governor-General of the Priamur hoped that the building of a good harbour at Nikolaevsk would ensure to the steamers the prosperity which is threatened by the opening of the Amur Railway, and also establish a large bean and grain industry on the Sungari and the Amur between Blagovyeshchensk, Nikolaevsk, and Kharbin.

Transbaikal.—The Transbaikal is sparsely populated. Cattle-farming is much more developed than agriculture. But the quality of the land is excellent. Spring corn is the chief cereal, forming 50 per cent. of the total crop. Then come wheat (16 per cent.) and oats (15 per cent.). Barley, buckwheat, and in places millet, flax, and hemp are also sown. But harvests are uncertain, and there are frequent local failures. Artificial irrigation is necessary in the principal agricultural districts, such as the Nerchinsk and Aksha, the natives using ditches for the purpose, but the irrigation should be systematized. In the Selenginsk district the sowing takes place early in May and the harvest in August, before the autumn frosts begin.

Potatoes are grown everywhere for local use, but other vegetables only on a small scale. Considering the high prices realized, it is surprising that they are not more cultivated.

Amur and Maritime Provinces.—Agriculture is at present the chief occupation in the Amur Province, but in the Maritime Province, in spite of the richness of the soil, the climate is not favourable to its development. There is little snow in winter. Hence the land freezes so deep that the subsoil remains frozen throughout the year. In summer the rainfall is heavy. Damp is the chief enemy. Clover and fodder-grass do not flourish, and there is much rot. Of some 567,000,000 acres only about 850,000 were sown in 1906. But the area under crops is steadily increasing, especially in the Ussuri region and on the Zeya-Bureya plain.

Amur Province.—At the time of the first Russian invasion, the Daurians on the Amur were great agriculturists. The soil along the railway, which freezes to a depth of 200 or 300 ft., consists of a sticky clay that only thaws for 3 ft. This is covered with coarse grass or scrubby vegetation. But the Zeya–Bureya plain is very fertile, and is now largely settled. Fields of corn extend as far as the eye can reach. Between 1907–10 an average of 220,892 tons of cereals was produced, and the output should increase rapidly.

The following are the statistics of the 1911 harvest :

	Tons.		*Tons.*
Oats	118,000	Beans	56
Wheat	110,200	Millet	1,200
Rye	2,500	Corn	19
Barley	980	Hemp and flax	140
Buckwheat	600	Potatoes	21,900

Potatoes do well, averaging 12 cwt. to the acre.

The Amur Province is the best market for agricultural machinery in eastern Siberia.

In 1910 there were 43 steam mills, 111 water-mills, and 60 windmills in the Amur Province. Blagovyeshchensk possessed 9 steam-mills. There were also 38 rice and groats mills in the Amur and Maritime Provinces.

The following table shows the flour produced in 1911 by the Blagovyeshchensk mills. Of the 100,806 tons of grain used, 37,250 came from the Amur Province, and 63,556 from Manchuria. The 1,310 tons of rye were all grown in the province.

Flour produced for :

	Wheat. *Tons.*	*Rye.* *Tons.*
Blagovyeshchensk	28,200	400
Mining districts	8,560	84
Zeya-Pristan	6,250	56
Commissary Department	6,500	620
Other markets	19,900	140
Total	69,410	1,300

In the same year Blagovyeshchensk imported from Manchuria 2,347 tons of wheat flour, and from Odessa and other places 305 tons.

In the short summer, carrots, potatoes, tomatoes, cabbages, even water-melons, and of course beets, do well almost everywhere on the newly-cleared land by the Amur Railway. The Chinese and Koreans grow most of the vegetables. Blagovyeshchensk is entirely supplied by the Chinese across the river, with whom the Russians cannot compete.

Maritime Province.—The mountainous character of the Maritime Province as a whole is not favourable to agriculture. The fertile territory lies to the south, round Nikolsk-Ussuriski and Lake Khanka, and in the Ussuri valley. The winter is long, with little snow, while an annual precipitation of 22 inches of rain falls between May and September. Fogs are common in the south, and many parts of the province suffer from floods. These are especially frequent on the Iman, while east of Lake Khanka in the Prikhankoisk region 500,000 acres of arable land are made useless by inundations in summer. It has been proposed that the district should be reclaimed by improving the outflow through the Sungacha. In 1910, 80,192 acres were damaged by frost, floods, insects, and other causes. The Russians had to learn from the Koreans how to protect their fields from floods and fogs. The Cossacks, who own large tracts of land, preferred to let it to these skilled farmers, whom the Government policy of Russifying the region is steadily driving out. It is impossible for a Russian peasant to compete with a Korean, not merely on account of the extraordinary skill of his intensive farming, but on account of his low standard of living.

The average size of a holding in the Maritime Province is about 15 acres. Means of communication are very defective, but the Government is building new roads. Farmers are also encouraged to sow soya beans, flax, hemp, and beets, for which the country is better suited than for grains. Indeed, the beets are so good that there is talk of starting a beet-sugar factory in Vladivostok. An agricultural school, the first in the Far East, was to be established at Nikolsk-Ussuriski, the centre of the flourishing farming district in the south.

The yield of the various crops in 1911 was as follows:

	Bushels.		*Tons.*
Wheat	2,197,249	Buckwheat	15,274
Rye	231,317	Millet	6,049
Barley	197,837	Soya beans	3,309
Oats	3,836,784	Flax and hemp	2,944

There are 115 steam-mills, 220 water-mills, and 139 windmills in the Maritime Province.

Until some attempt at drainage is made, and the old coarse grass is removed, the hay in the Ussuri district will always be rank. It often smells so disagreeable that cattle refuse to eat it.

Potatoes are a very important crop. Vegetables do well and sell well, but the Chinese and Koreans reap most of the profits.

Considerable success has been obtained in fruit-growing, and attempts are being made to establish it on a business basis. At Barabash an apple is said to have been produced not inferior to the Californian varieties. It is hoped that in time the large imports of fruit, especially from Japan, will become unnecessary. Grapes of a poor quality grow wild in the southern Ussuri district. The inhabitants make wine from them for local use.

Sakhalin. Wheat, barley, oats, spring corn, and even winter rye are sown, mostly in small patches by ex-convicts. Three- to four-fold is a good harvest. Vegetables, notably potatoes and cabbages, do pretty well. The soil is good in places.

Yakutsk and the North-east.—A recent commission has decided that agriculture could be advantageously carried on in the Yakutsk Province. Hitherto it has proved most successful in the Lena valley and on the Olekma and Aldan. The flourishing Skoptsi colonies near Olekminsk and Yakutsk grow wheat, summer rye, barley, and oats. These Skoptsi import modern agricultural machinery, and own two or three steam flour-mills. Most of the Yakuts grow wheat. In fact, grain already ripens at about lat. 64° N. Barley ripens in

71 days, spring corn in 92, oats in 82, and wheat in 76 days. But the harvest is very moderate.

Vegetable-growing in a small way is carried on as far north as the Kolima and the Verkhoyansk regions. Barley sometimes ripens near Verkhoyansk. The Skoptsi of Markha, near Yakutsk, are remarkably successful with vegetables, especially potatoes and cabbages. But the natives of the tundra are too busy fishing in summer to waste their energies on occupations so unremunerative as agriculture and vegetable-growing. The bleak shores of the Sea of Okhotsk are very unfavourable to agriculture. Even potatoes will not ripen near the mouth of the Uda, though barley and excellent vegetables are grown inland round Udski-Ostrog. Yet cabbages and even cauliflowers do well in the more sheltered region of Yamsk. In Kamchatka the rich black earth and comparative dryness of the soil between Verkhne-Kamchatsk and Klyuchevskaya make agriculture possible in the valley of the Kamchatka, and barley and vegetables ripen round Petropavlovsk. Indeed, vegetables such as potatoes, beets, carrots, radishes, and cabbages are grown in most of the villages of the peninsula. In the Petropavlovsk district over 100 acres are sown with vegetables. Potatoes and turnips have ripened at Siktyakh, on the Lena, and there are several small kitchen gardens at Markovo on the Anadir. The example of Alaska shows that there is no reason why the growing of vegetables should not be widely extended in these regions.

Opium and Tobacco

The Ussuri region is very favourable to the poppy, and it used to send a quantity of opium to China. Russians are not addicted to the drug, but the Cossacks, who are not good agriculturists, readily let their land to the Chinese grower for £10 an acre. They are thus enabled to live in idleness, and they rapidly become demoralized. Not only does the poppy exhaust the ground, but it diminishes the output of beehives in the neighbourhood by 75 per cent. Moreover, it attracts Chinese of a very undesirable type. In 1911 some 10,000 acres were under poppy, and much more is probably grown illicitly in the remoter districts. Its sale, except as a drug, is illegal in Russia, but there is a large contraband Chinese traffic. Indeed, so alarmingly has the trade increased, that the growing of the poppy is to be prohibited by law. In 1913 the crops were destroyed by order of the Government.

Some tobacco is grown in the Yeniseisk and Irkutsk Provinces. About 600 acres were under tobacco in the Maritime Province in 1911.

Cedar-nuts

Cedar-nuts, which give good oil, are gathered for eating in the Amur Province, and in much larger quantities on the upper Yenisei and the upper courses of the tributaries of the Angara. In the Sayansk taiga, especially, the industry is organized on a commercial basis. September and October are the months for collecting. Wooden mortars are used for husking the cones. The forests lie in the heart of the Sayansk taiga. In a good season, which occurs every 4–5 years, 800 tons are sent to Krasnoyarsk from the Yenisei.

Crops of Siberia and Steppes

	1913.	*Mean crop.* (1908–12).
	Tons.	*Tons.*
Wheat	3,215,920	2,069,250
Barley	254,460	158,110
Oats	1,899,310	1,239,240
Rye	888,020	739,090
Hay	11,549,860	
	1914.	
Buckwheat	58,840	
Millet	202,500	
Maize	5,870	
Leguminous crops	31,930	
Potatoes	1,637,500	

AREA UNDER CROPS IN 1911

Province.	*Total Cultivated Area.*	*Cereals.*	*Potatoes & Leguminous Crops.*	*Flax.*	*Hemp.*
	Acres.	*Acres.*	*Acres.*	*Acres.*	*Acres.*
Tobolsk . .	3,821,500	3,679,060	64,416	51,529	26,495
Tomsk . .	7,183,380	6,822,341	174,250	91,950	94,840
Akmolinsk . .	1,801,730	1,687,035	40,073	69,473	5,146
Semipalatinsk .	628,200	617,190	7,680	2,400	930
Irkutsk . .	993,300	960,600	23,500	1,300	7,900
Yeniseisk . .	1,223,000	1,159,100	34,840	9,280	19,780
Transbaikal .	929,700	907,400	20,700		1,600
Yakutsk . .	33,800	32,600	1,200		
Amur . .	698,500	683,940	13,200	400	960
Primorsk . .	656,200	604,700	40,500	8,130	2,870
Sakhalin . .	5,700	4,560	1,140		
Total .	17,975,010	17,158,526	421,499	234,462	160,521

CHAPTER X

LIVE-STOCK

Domestic animals—Apiculture—Dairy industry

Domestic Animals

In western Siberia the raising of live-stock is a great occupation of the inhabitants, though more to meet their own needs than for the purposes of trade with other nations. In eastern Siberia, especially with regard to horses and cattle, it is better developed than agriculture, but is still far from sufficing for the needs of the country. The figures for the principal live-stock in 1911 were as follows :

Province.	*Horses.*	*Cattle.*	*Sheep (coarse wool)*
Tobolsk	749,672	900,201	624,215
Tomsk	2,173,733	2,469,049	2,403,724
Akmolinsk	1,064,229	1,289,061	2,064,462
Semipalatinsk	957,926	856,830	2,792,699
Yeniseisk	494,157	500,139	724,004
Yakutsk	88,138	241,674	135
Irkutsk	287,518	354,622	203,518
Amur	101,206	75,231	11,836
Transbaikal	591,588	1,003,145	1,021,967
Primorsk	109,516	171,618	4,640
Sakhalin	1,797	4,315	
	6,619,480	7,865,885	9,851,200

Province.	*Sheep (fine fleece)*	*Goats.*	*Pigs.*
Tobolsk	3,251	20,713	154,786
Tomsk	23,118	64,168	508,253
Akmolinsk	27,125	133,185	45,359
Semipalatinsk	13,251		8,063
Yeniseisk	4,775	7,472	123,307
Yakutsk			217
Irkutsk		29,468	79,452
Amur		83	48,200
Transbaikal		103,082	119,366
Primorsk		200	91,187
Sakhalin		74	1,400
	71,520	358,445	1,179,590

Horses predominate among the Kirghiz, who breed them for transport, meat, and *kumis*, which is manufactured from their milk : the Kirghiz ride wherever they go, however short the distance. When they change from a nomadic to a settled life, the number of their horses decreases and that of their cattle increases. The tribes of the Altai largely breed horses; from their mares' milk a spirit called *terasum* is prepared. In the rest of Siberia the horses are bred mainly for farm-work or to act as post-horses.

The chief breeds in Siberia are the Kuznetsk and Kirghiz. In the south of Tobolsk the native horse is interbred with the Kirghiz variety. This cross-breed is of extraordinary speed and staying power. The horses in Tomsk are bigger; they are not so swift, but can carry great weights. The Siberian horse is usually small, easily satisfied as to food and water, and can endure heat and cold. It is fast, but not extremely strong : its normal load does not exceed from 720 to 900 lb. Only the superior sort of dray-horses draw 1,000 to 1,080 lb. (or, for a short distance, 1,260). There is but little breeding with English and American horses, except for the carriage-horses of very rich men. But high-stepping horses thus bred can be seen at certain places, e. g. Tyumen and Tomsk.

There are certain breeding-places, especially in the Tomsk Government, where stallions are kept; their number was much increased in 1912. Besides these there are stud farms in the Tomsk, Tobolsk, and Semipalatinsk Provinces. Horse-shows have been introduced in the towns of Tomsk and Barnaul and the village of Bryukhanovo, in the Tomsk Government.

In eastern Siberia horse-breeding is an important industry. The Transbaikal horse is the best known and most popular in the southern districts. The Cossacks of the Amur prefer it to any other, and it is the only breed used in Sakhalin. It is small, thin, hardy, and well suited to endure a rough life. It is 12 or 13 hands high, can draw a load of 1,000 lbs., and cover any distance at 40 miles a day in a troika. So light is the snowfall in the Transbaikal and Amur Provinces that it can graze all the year round in the open. The *vostretz* grass (meadow-grass) of the Transbaikal, which affords almost better food in winter than in summer, keeps the horses in condition there. In the Amur Province, where there is no such grass, they get thin in winter, though they quickly fatten again in spring. But the breed degenerates rapidly there. The Government has recently started stud farms with good stallions in the Transbaikal and Yakutsk Provinces in order to improve the breed. In the Transbaikal the average price of a horse is about 5 guineas, at Blagovyeshchensk from 10 to 20 guineas.

In the Yakutsk Province the ugly little Yakut horse, with its shaggy coat, displays astonishing endurance. It often lives out of doors in winter, and is even used within the Arctic circle.

Cattle.—The Siberiaks are much more careless about their cattle than the emigrants, for they allow them to remain out in the winter without any shelter except such as is given them by their coats, which are as a rule very thick. Cattle-raising is especially developed in the Tyukalinsk district of Tobolsk and in the Kainsk steppe and about the Chulim in Tomsk. Cattle have increased in value ; formerly they were worth about £1 each ; latterly they could command from £2 to £4. The local prices for meat range low; so it is more profitable, when possible, to use them for dairy purposes. Consequently there has been a decline in the hide and meat trades. The export of cattle themselves is inconsiderable—in 1911, 65,000 head (£250,000). The great commercial centres of the cattle trade are Petropavlovsk and Omsk (where the railway crosses the Ishim and the Irtish). From the former meat is exported to European Russia at the rate of 30,000,000 tons a year. Further west there is not enough cold storage for trade, but some firms are establishing cold storage plants along the railway. In the steppes cattle are bred for meat, as well as for dairy purposes, but the meat is mostly for local consumption, and of little economic value, though a certain amount is exported frozen in winter to Petrograd, Moscow, and the far east from the region between Petropavlovsk and Novo-Nikolaevsk; there is no export in the summer. Some time ago a German company was said to be organizing the export of Siberian meat, especially veal, to Berlin.

Both the Yeniseisk and Irkutsk Provinces are poorly off for cattle. But stock-raising is the chief occupation of the nomads of Minusinsk, Achinsk, and Turukhansk, and of the natives of the Balagansk and Verkholensk regions. The Minusinsk district alone has enough for its own needs. In southern Yeniseisk each peasant possesses, on an average, 2 horses, 5 cattle, and 10 sheep. All the oats and hay are kept for food for the stock in the winter. Large herds of the fine, fleshy Soyot cattle enter the Irkutsk Province every year through the Sayansk Range for the supply of meat to Siberia. As many as 30,000 cattle are said to reach Irkutsk by a single track. The animals are kept in quarantine for a fortnight and are medically examined before crossing the frontier.

In Transbaikal cattle-raising is very successful, especially among the nomad tribes. Thanks to the *vostretz* grass, it should be capable of considerable development. There is even a co-operative society, with its centre at Chita, which engages in purchasing cattle for the army. As many as 150,000 hides are annually exported from this province. They are used to cover tea-chests. Owing to the rough life the effects of attempts to improve the breed have only become visible in the last few generations. The average price of a cow is £3 15*s*.; of a pair of bullocks, £10 15*s*.

In the Amur and Maritime Provinces the cattle are Manchurian or Korean. As they are never milked at home, they only give milk while the calf is with them. In Blagovyeshchensk cows of local breed cost from £8 5*s*. to £12 10*s*; bulls from £8 6*s*. 8*d*. to £12 10*s*. Some authorities believe that the future of the Amur Province lies in cattle-breeding rather than in agriculture, and vigorous efforts are to be made to encourage it. At present the greater part of the meat supply comes from Manchuria. In 1907, 5,000 head of Manchurian cattle and sheep were imported. The chief fairs are at

Blagovyeshchensk, Khabarovsk, Nikolsk-Ussuriski, and Vladivostok. Australian meat has been imported into Vladivostok in small quantities.

The cattle are of small breed; the average live weight of cows is 510 to 650 lb., of bulls 800 to 930 lb.; the Altai breed is bigger. On account of the rigorous climate of Siberia it is little use to import foreign breeds. A small number of Simmenthal and Allhausen cattle have been purchased at Moscow and introduced into Siberia, but only by the well-to-do. Cattle as a rule are free from tuberculosis, but suffer from Siberian plague and from foot and mouth disease (*yashchur*). Quarantine regulations have practically extinguished rinderpest, which in certain years, such as 1884, wrought great disaster. Despite the primitive nature of their keep, the Siberian cattle stand the climate well.

Sheep.—The ordinary Siberian sheep is of a poor breed; it yields little meat, very little tallow, and inferior wool. The Kirghiz sheep vary in colour; they are grey, white, black, and sometimes red; the Kirghiz prefer the grey and white, as their wool fetches a better price in the markets. They are shorn twice a year, yielding from four to six pounds of unwashed wool, which is largely used for felt, both for local needs and in the factories of western Siberia. The Kirghiz sheep in winter cannot get at their food through the snow; so they follow in the wake of the horses and cattle, and eat what pasture is left for them. One type of Kirghiz sheep has a thick pad of fat on its rump, which affords much tallow. Tallow factories are common in the western Siberian towns, but there is not enough to meet the local needs, and tallow is imported.

Since 1900 a breed of fine-fleece merino sheep has been introduced; it is of Spanish origin and was reared in the Crimea; it prospers in the Semipalatinsk and Akmolinsk Governments, and in the Altai in the Zmyeinogorsk district. The average yield of unwashed wool is 13 lb. to 15 lb., the best rams yielding up to 36 lb., the best ewes up to 22½ lb. The wool sells at Kharkov from 16*s*. to £1 4*s*. the pud. The sheep are brought by considerably reduced tariffs to the steppe and Altai regions. The wool exported to European Russia in 1911 was worth £400,000. A considerable quantity of the wool goes to Irbit and thence to Nizhne-Novgorod and the interior of European Russia by rail and water, or else by the Irtish to the Perm Railway, and so by the Kama and Volga to the centre of Russia. In old days the sheep that found no place in the fresh-meat market were sent to the *salgani* and salted for the winter, and sold for as little as 1*d.* or ½*d.* the lb.; now only a small amount of meat goes to the *salgani*.

Sheep-breeding seems likely to become an important industry in the Yeniseisk and Irkutsk Provinces, now that the merino sheep has been acclimatized in central Siberia. The plains of the Yenisei have been found to be suitable for sheep-farming on a large scale, and it is probable that the numbers are already far greater than in 1911. The soil yields excellent forage and in a good season sufficient hay can be obtained for several years. The cost of transferring flocks from Russia is heavy, but the Government is providing subsidies for this purpose and for the development of new sheep-farms. The sheep are killed for their wool and their tallow. There is as yet no regular demand for mutton. Indeed an elaborate system of refrigerators would have to be organized before it would be possible to find and supply a regular market with meat.

In Transbaikal the sheep are large, with an average height of over 2 ft. The wool is coarse, thick, and of medium length. The average price is £1 12*s.* 2*d.* In the Amur Province the absence of dry pasturage is unfavourable to sheep. The few that are bred are Mongolian.

Goats are bred by the Kirghiz, Cossacks, and Russian peasants. The industry of preparing goat's hair coverings is developing among the emigrants in Turgai. Large numbers of goats are kept by the natives of Transbaikal. In 1914 there were 114,105 there.

Pigs.—The Russian pig is a strong animal, and stands the

discomforts of temperature and crowding well. It is only the Russian population who keep pigs east of the Ural; the Kirghiz, being Mohammedans, are not allowed to eat or keep them. Development of the bacon industry promises well, because the growth of dairy-farming has meant abundant supplies of butter-milk for pig food. The amount of bacon exported went up from 700 tons in 1908 to 4,800 tons in 1911. There is a proposal for a British company to acquire the waste lands along the banks of the Ob, and to place them under grass-cultivation for breeding bacon-pigs, so that there should be ultimately an enormous bacon-export by the Kara Sea. Kurgan contains a sausage factory, and there are at least two bacon factories in Siberia.

In Yeniseisk, Irkutsk, and Transbaikal pig-breeding is carried on extensively. In Transbaikal the animals are Manchurian and of poor quality. They are small and lean, but they increase rapidly. The breed is being improved by imported swine. The average price is £1 12*s.* 2*d.* Pigs are not numerous in the Amur Province, but in the Primorsk the pig-industry is considerably developed, especially on the Ussuri. Greater care might make it more successful than it has proved hitherto. Here also the breed is Manchurian.

Camels are bred by the Kirghiz, and in a few cases by Russian emigrants: those with single humps are larger, and supposed to be stronger than those with two. The camels are better cared for by their masters than the cattle are; they are very sensitive to cold, and so winter forage is stored for them. They are used for transport purposes and also for their milk and hair, which is obtained once a year in the spring, giving an average of 12 lb. per camel.

The camels in the Transbaikal, of which there were 10,992 in 1911, are strong and of great endurance. A pair of them can carry 20–25 cwt. The average price is about £10 15*s.*

Reindeer are found in great numbers in the north of Tobolsk Government and to some extent in the Narim district of Tomsk. It is very difficult to obtain any idea of their numbers, but they have been estimated at 515,000 in Tobolsk, and 2,000 in Tomsk. To the Ostyak and Samoyede who possess them, the reindeer means everything—milk, meat, clothing, travelling, and traction.

There are small herds of reindeer in the Amur and Irkutsk Provinces. In 1906 there were said to be 515,000 reindeer in the Yeniseisk Province, principally in the Turukhansk region, 95,360 reindeer in the Yakutsk Province, 80 per cent. of them being in the Verkhoyansk and Kolima districts. In the Kamchatka Province, including the Chukchee Peninsula, there were 287,000 reindeer. These numbers can only be approximate. The Chukchee have the largest herds of reindeer in the world.

In the north reindeer and sledge dogs are the only domestic animals.

Maral deer are kept in the Altai, along the upper Irtish, and in the southern regions of the Yeniseisk and Irkutsk Provinces. They are a kind of wapiti, which are kept in special farms (*maralniki*). Their horns, cut off in the velvet, are sold to the Chinese, who extract from it a drug (*panty*), much esteemed by them, but of very doubtful efficacy. The price of horns fluctuates very much, depending on the state of the markets : it commonly varies from 14*s*. to £1 per lb. But those horns which are sold with the frontal bone are considerably dearer—£10 or even occasionally £20 per lb.—as it is then necessary to kill the animal. As the horns attain a weight of 10 lb. and the average weight is 5 or 6 lb., and the keep costs practically nothing, it can be seen that this form of deer-breeding is very profitable. In Transbaikal the *izyubr* or wapiti (*Cervus canadensis*), which closely resembles the maral, is bred instead, while in the Primorsk, in addition to the izyubr, the aksis, the most valuable of all these deer, is found. The horns of the aksis fetch from £3 to £3 10*s*. per lb. Of the 10,000 deer of this type in *maralniki* in Asiatic Russia in 1911, 3,125 were in the Maritime Province, especially near Olgi Bay and the Suchan River. The industry is likely to develop still further. The kabargi or musk-deer (*Moschus moschiferus*) is also being bred commercially ; it is highly valued by the Chinese for the medicinal properties of its musk and its horns.

Poultry, &c.—There is a certain export of eggs from Kurgan. Owing to the presence of many lakes and ponds, duck- and goose-breeding might have considerable development, but it takes a long time to start. Siberian geese are comparatively small.

Apiculture

This is a very old industry in Russia, recorded as long ago as the eleventh century; the Russian princes used to levy tribute of wax and honey on their conquered subjects. The bee has had almost a sacred character because of the wax candles used in religious worship. The decay of the industry is said to be due to the drying up of the steppes, and in some parts to the destruction of the forests; but there are parts of western Siberia in which it is very active, expecially the districts of Kuznetsk, Biisk, Zmyeinogorsk, and Ust-Kamenogorsk; there is also some beekeeping as far north as the district of Tomsk, and among the Urals, where there are *artels* for beekeeping, e. g. at Verkhne-Tagilski. In the Achinsk and Minusinsk districts, in the south of Yeniseisk, there are over 45,000 hives. The Little Russians have here introduced the latest methods. Apiculture in the Amur Province is on a small scale, but growing rapidly in importance. It is concentrated along the Byela, the Khara, and the Zavitaya, where 92 villages carry it on on commercial lines with 2,757 hives. In 1911 44,345 lb. of honey and 4,434 lb. of wax were sold. The climate and vegetation of the Amur Province are very favourable to apiculture. Its slow growth is due to the fact that the peasants find other work more profitable. Bee-keeping is also very successfully carried on in the Maritime Province, notably round Nikolsk-Ussuriski.

Besides *artels* in certain places other help is given to beekeepers: a considerable number of expert instructors is available, and local government bodies sell bee-culture equipments at reduced rates, where desirable. It is estimated that the collective output of Siberia is about one-fifth of the output of the whole Russian Empire. The most recent figures available are those for (*a*) private owners in 1910, and (*b*) peasants in 1908.

	Number of hives.	*Honey.*	*Average price per pud.*			*Wax.*	*Average price per pud.*		
		(*tons*).	£	*s.*	*d.*	(*tons*).	£	*s.*	*d.*
(*a*) Siberia .	682,878	1,765		16	6	197	2	0	8
" Steppes .	75,296	154		17	2	20½	2	1	9
(*b*) Siberia .	640,892	1,657		12	1	183		1	3
Central Asia .	71,686	313¾	1	0	3	31		2	2

Dairy Industry

The dairy industry has developed with more astonishing rapidity than any other undertaking in Siberia. Before 1893 only *toplennoe* (boiled or melted butter) was manufactured in Siberia ; it sold at an average of 12*s.* per pud, and required 32 puds of milk to produce one pud, the average price of milk being 4½*d.* or 4¾*d.* per pud. It is still made by the Russian peasants but not exported. In 1893 the first dairy farm was begun near Tyumen by the English wife of a Russian. The new article sold at from £1 to £1 4*s.* the pud, and required 20 or 22 puds of milk for one pud. In the course of ten years the amount developed, under government patronage, to 32,000 tons, worth about £2,500,000. The growth of the industry was so astonishing that between 1900 and 1902 the number of dairies had increased by 91 per cent. Government encouraged it by granting loans, for which the live-stock of the peasants was sufficient security, and subsidies to village communities for establishing dairies. In 1903 an additional £200,000 was granted for starting dairies on the *artel* system, by which several peasants distributed the proceeds in proportion to the amount of milk provided ; it becomes a general or public dairy when the group is extended so as to include the whole village community. Further, the Government established technical dairy schools at Kurgan, Omsk, Kainsk, Barnaul, and Zmyeinogorsk, a central laboratory at Tomsk, and local laboratories in five other centres, and refrigerating stores began to be erected along the railway.

Butter is small in bulk and can be profitably exported. The export trade, which began in 1897, is mainly in the hands of firms in Moscow, hitherto largely managed by Germans and Danes. Lately more interest has been taken by British commerce in this industry, but only one or two British firms have established offices in Siberia, and only one British house has regular offices in its own name at the principal centres of export. Siberian cheese also finds a ready market at home, but the better grades are also sold throughout Europe.

Siberian milk contains an exceptionally high proportion of fat. The pasture is rich, and the cattle eat much and drink little. Owing to its richness the average yield is 1 lb. of butter to 20·05 lb. of milk, whereas in Denmark it is to 28 lb. The Siberian cow is long lived; its meat is inferior, so that most are kept for dairy purposes; fodder cannot be exported at a profit, and will therefore continue to be cheap in the country. Cattle are now fed scientifically on preserved fodder. A cow brings in from 14*s*. to £1 1*s*. a month in accordance with the season.

The Government aim at having an *artel* dairy in each village; any one of average industry and intelligence after three months' training at a dairy school is competent to be engaged by an *artel* to direct their labour. There are instructors with about ten dairies under them, who have usually been Germans or Danes. Most of the dairies work with hand- or horse-power, steam being less applicable in out-of-the-way places. Pasteurization of milk is being introduced, and cement floors in dairies are to be obligatory. The dairies purchase milk from neighbouring farmers and cream from those more distant. Many farmers churn their own butter and sell it to the dairies to be rewashed, tested, classified, packed and sent off to the market-centres. The cost of milk at the dairies is 4*d*. a gallon; the price obtained for butter varies from 9½*d*. to 8½*d*. a pound. In 1910 15 butter factories produced up to 7,500 lb.; 30 between 7,500 and 12,500 lb.; 15 up to

18,000 lb.; 11 up to 25,000 lb.; 8 up to 40,000 lb. In 1912 there were 1,060 dairies in Tobolsk, and 2,042 in Tomsk. Of this total of 3,102, there were 1,784 in private hands, and 1,318 managed by *artels* (42·5 per cent. of the whole).

Everything is done to help the export trade, which is directed to the Baltic ports by butter trains, which take precedence of all other goods traffic. They start from Novo-Nikolaevsk during the summer (especially in June and July) ten to fourteen times a week. The ice-trucks have a carrying capacity of from $7\frac{1}{4}$ tons to $22\frac{1}{2}$ tons. The train is made up at the various butter-transit centres with ready-loaded trucks as far as Chelyabinsk, and reaches the number of twenty-five trucks. Among the chief centres of the trade are Omsk, which is a sort of clearing-house for exportation of butter, and Kurgan, which is the main distributing-point for home and foreign markets, and where all butter exported is subject to a preliminary investigation by state officials, members of the Agronomical Organization. On arrival at the ports, especially Riga (and Windau before the war), it is loaded on special refrigerators, and so on to the steamers which convey it to London, Hamburg, Hull, and Copenhagen, of which the last named does a considerable amount of re-exportation. The sea-freight to British ports costs about 5*d.* per pud. There are extensive refrigerators at the chief loading stations.

This is the main route of the butter trade, but the Kurgan Farmers' Association has been pressing the importance of pushing the trade in south European Russia, and there is a great opening for it in the Far East, which depends on tinned Canadian milk and tinned Australian butter. It is said that Siberian butter actually improves by being kept in cold storage, even up to six months.

A fewfigures will show the rapid development of the industry and its present proportions. Starting from nothing in 1893, the export of butter in 1903 was 35,225 tons, the principal centres being Kurgan, which exported 8,227 tons, and Novo-Nikolaevsk, through which came the Altai butter, which exported $8,066\frac{1}{3}$ tons. In 1912 there was a great development;

the transport by the Siberian railway increased by 9·9 per cent. between April and October on the transport during the same period the year before. To eastern Siberia 7,800 tons were exported as compared with 5,000 tons the year before. A larger proportion of what was sent west was sold in the interior markets of Russia. The exports to Britain increased; those to Germany and Denmark decreased. There were exported from Siberia, chiefly from Tobolsk and Tomsk Governments:

in 1910, 63,814 tons, worth £5,836,900;
in 1911, 70,377 tons, worth £6,553,800;
in 1912, 71,934 tons, worth £6,800,000.

In 1913 the amount exported was 76,000 tons, worth £7,160,000, and for home consumption 71,500 tons, a total of 147,500. The war caused a great decrease owing to abnormal conditions and the closing of the Baltic ports, but even in 1914 there went from Barnaul 24,130 tons, and from Omsk 18,252 tons.

Dairy-farming is of much less importance in eastern Siberia. Yeniseisk only supplies 1·2 per cent. of the Siberian dairy produce on the market. In Transbaikal the cattle are small and yield little milk; this is especially true of the Buryat cattle, Mongolian in origin.

Dairy-farming is slowly increasing in the Maritime Province. In 1911 this province only contained five co-operative dairies, but in 1913 there were eighteen. In 1911 local butter cost from 1*s*. 11*d*. to 2*s*. 2*d*. a lb., but in 1913 it was only 1*s*. 2*d*. Nearly all the dairy produce is the work of the four summer months. The local dealers often have contracts with west Siberian firms. Hence the butter fails to become known locally. Proposals have been made for building cold-storage plants at Vladivostok.

Cows have been known to give milk in Sredne-Kolimsk and Verkhoyansk, and even further north, but cattle-raising can never be profitable in these regions.

CHAPTER XI

TIMBER INDUSTRY

Timber and Timber Trade—Sawmills—Wood Industries

Timber and Timber Trade

In comparison with the vast area of Siberia covered by forests, the timber industry is inconsiderable. The absence of communications, the fact that much of the timber grows in inaccessible parts, the destruction of the forests in those parts, like the banks of the Ob and Irtish, whence it is most easily transported, the bulk of timber, and therefore the difficulty of transporting it, are among the causes which have produced such a discrepancy between the productivity of the land and the economic results.

Extent of Forests.—It is estimated that there are altogether 853,000,000 acres of forest in Asiatic Russia, of which the largest part is unexplored, while a considerable portion towards the north consists of valueless trees. The State is the owner of 642,000,000 acres, of which 39 per cent. is classified as rich forest soil. This is a small proportion, for 80 per cent. is so classified in European Russia. Next to the State the Crown is the principal owner with 54,000,000 acres, principally in the 'cabinet' estates of the Altai.[1] The State forests are divided into forestry districts, which are sub-divided into allotments.

Of these State forests of Asiatic Russia up to January 1, 1912, only 8,300,871 acres were organized, and only 67,924,917 acres are investigated, these two amounting to something like 12 per cent. of the whole. In 1912 and 1913 another

[1] No information is available as to the disposition of the Crown estates since the revolution.

44,075,900 acres were organized and investigated, and it was hoped that by the beginning of 1915 nearly a quarter of the whole area of State forests would have been thus dealt with. The average Siberian forestry district exceeds the wooded area of all Great Britain (3,037,500 acres). The average allotment is 311,850 acres, but this computation includes a great part of the northern district, which is practically valueless. No wood of economic value is obtained north of lat. 60° N., though larches grow to lat. 68° N. on the east side of the Urals. The height and diameter of the northerly trees are small; they grow very slowly, and their wood is weak and poor. Taller and better trunks are only to be found at some isolated, dry places, but, unless trees grow conveniently close to one another, they are not of much commercial value.

Trade.—The Siberian timber trade, as a whole, is disappointing. In 1911 in the whole of Asiatic Russia 1,800,750,000 cubic feet of timber were assigned for sale, but only 10 per cent. was sold. From the State forests of Asiatic Russia in 1910 were realized £423,400; in 1911, £407,900; in 1912, £425,900. The value of each desyatin (2·7 acres) was roughly 1*s.* 10½*d.* in Akmolinsk; 1¼*d.* in Tobolsk; 1½*d.* or 1¾*d.* in Tomsk. As regards the State forests of Siberia only, in 1911 there were 558,002,916 acres, of which 211,087,351 were worked directly by the Government. The birch-bark, bast, &c., produced were about 1,260 tons. The total amount of produced material was 126,662,011 cubic feet. The gross receipts were £273,293, and the net profit £60,994.

Administration.—Paid forest guards are appointed to look after the forests of western Siberia, and the peasants are supposed to see to the forests put at their disposal. The administration only allows that part of the forest to be cut each year which is specially assigned, and control is exercised over the raftage and steamer-wharves. The Forestry Administration is undermanned. A law was passed in 1889 that those engaged in the timber trade should, under payment of a deposit, replant the forest land laid bare. But this is seldom done, most deposits are forfeited, and only 14 per cent. is replanted. Forest fires are a frequent calamity ; they are caused mainly by the burning of the grass in spring and by sparks from engines. The latter cause, coupled with the hewing of wood to meet the needs of the railway, means that there is never much forest in the neighbourhood of the lines. Severe storms also do much damage where the wood is thin.

The Western Provinces.—The most recent figures for the State forests in the western provinces are as follows :

	Forest Districts.	*Allot-ments.*	*Area : acres.*	*Suitable Forest Soil : acres.*
Tobolsk . .	33	497	170,828,817	45,563,156
Tomsk . . .	27	190	91,837,876	22,047,425
Semipalatinsk .	11	190	9,299,254	5,818,675
Akmolinsk . .	15	302	6,130,293	1,119,344

In the Tobolsk Government the best timber is thought to be in the valleys of the Tavda, Tura, and Pelim ; estimates differ considerably, but the more sanguine assign to these valleys about 27,000,000 acres of good timber. In Tomsk the forest belt is in the north part ; the south part is half forestless, but the Kuznetsk and Mariinsk districts and the Altai mountain region are forested, though in many parts the forest is not at all dense. In the mountainous parts it is very difficult to cut the timber because of the precipitous places in which it grows. When felled, it is apt to fall into ravines below, not only being lost itself, but smashing other trees in its fall. There is no possibility of raftage on the rivers there because of their swiftness. Some timber is transported by camels across the Kirghiz steppe. But on the steppes and in considerable parts of the Governments of Tomsk and Tobolsk, the amount of forest is not enough for local needs.

Timber trees in Western Siberia.—The principal woods of Western Siberia are white cedar or cembra pine, pine, spruce, fir, larch, oak, ash, and birch (see Chapter III). Of these white cedar and larch are most valued for building, larch being especially used for boat-building, also for beams and telegraph poles. Birch is used for fuel and building, and the birch spinneys in the neighbourhood of towns rapidly disappear. The roofs of the houses of peasant and native are made of birch bark. Siberian cedar is of value because of its softness ; it can be used for certain kinds of furniture and pattern work, and it is worth exporting despite the cost.

Timber Trade in the West.—The most important place for the timber export trade in western Siberia is Tyumen ; an enterprising sawmill proprietor from Arkhangel used to send timber to Kotlas by rail ; thence it went to Arkhangel by steamer and so to London. The wood suffers from 'blue mould' on the voyage, and so does not fetch as high a price in England as would be anticipated, and the transport is very dear. It is difficult during the short summer to bring pine-trunks to the saw so dry that they avoid turning blue. Only first-class timber can be exported because of the cost of transport to Arkhangel and Petrograd by rail, and the export by sea by the Kara Sea route is negligible.

Eastern Provinces.—The forest areas of eastern Siberia have been very imperfectly surveyed. Little reliance can be placed upon official estimates, for there is a general tendency to exaggerate the wood resources of the country. Thus in 1908 a commission gave the wooded area of the Yakutsk Province as 540,000,000 acres. Such an estimate must include very large treeless spaces. In Yeniseisk, Irkutsk, Transbaikal, and Yakutsk the absence of suitable waterways makes the development of a lumbering industry out of the question, unless the northern sea-route from the mouth of the Yenisei can be opened to regular navigation. There is, however, considerable local consumption. The supplies of fire-wood in the immediate neighbourhood of the large towns and the principal mining-centres are, to a great extent, exhausted. Wood has to be rafted down the rivers, notably the Yenisei and the Angara. The building of the railway has greatly increased consumption, and the price of firewood has doubled in many places. Larch is the tree most used. Very fine woods of it are still found on the Yenisei, especially on its middle reaches. The Sayansk region, where trees grow at an altitude of 3,700 feet, has particularly fine forests, the taiga being almost impenetrable in many places.

The wood in the north and north-east is, as a rule, of poor quality. There is good timber, mainly larch, fir, spruce, and birch, in Sakhalin, but there is no suitable harbour for exporting it. Nor would it be possible to find a profitable market for the timber in the valley of the River Kamchatka, the only place in the Kamchatka Peninsula where it does well. A certain quantity of timber and firewood is annually rafted down the Lena, and a little also down the Kolima and Yana. In spite of the wide extent of the forests, the price is high owing to the scarcity of labour and the exorbitant rate of wages.

Lumber Industry in the East.—The Siberian lumber industry is at present confined to the Amur and Maritime Provinces. It is still in its infancy. The extent of the area under timber is uncertain, and it is impossible to give exact figures, as much of it is unsurveyed. A report for 1913 gives the total area of forest land in the far east as 110,052,000 acres, of which 30,840,000 are in the Amur Province and 71,463,000 in the Primorsk. In the Amur Province five-sixths of this area is Government property, and most of the rest belongs to the Cossacks; in the Maritime ten-elevenths of the area belongs to the Government. The forests cannot compare with those of North America in density. They generally lie along the sea-coast or in the river-valleys. The best timber districts in the Maritime Province are Nikolaevsk, with 33,000,000 acres, Khabarovsk with 14,245,000 acres, and the lower Amur with 12,150,000 acres.

Concessions.—The government forests are under the Department of Domains, with head-quarters at Khabarovsk. Areas that are for sale are knocked down to the highest bidder, who is obliged to deposit the royalty on the number of trees to be felled for one year. Felling tickets to cut small quantities of timber may be obtained from local authorities.

Hitherto forest grants have been limited to four years, but the term will almost certainly be extended to twelve, as a concession of four years was found too short for profitable working.

Royalties.—A royalty is levied on all timber. It is calculated on the cubic measurement of the logs sold, and varies according to the size and kind of timber. Formerly concessions lying more than from 6½ to 10 miles from a railway, a navigable river, or a bay, paid at a lower rate. Now, however, the tax is uniform, and it is hoped that this change will help to extend the range of forest-work and thus increase the revenue, which fell considerably short of the expenditure. But the absence of roads makes it unprofitable to fell timber at any distance from the rivers. Owing to the high railway rates only 10 or 12 per cent. of this can be profitably exported.

The Rivers and the Timber Trade.—The Amur and its tributaries provide an ideal artery for timber-rafts during the summer months, and when the suggested harbour improvements are carried out at Nikolaevsk, it should afford a good outlet for the trade in spite of the shortness of the open season. Export on a large scale is at present hampered by the poor harbours and the length of time during which the rivers are frozen, as well as by the difficulty of procuring labour.

Present State of the Trade.—So unfavourable were the economic and labour conditions in 1912 that only 12 out of 106 concessions were taken up, and only 17 firms instead of the usual 20 or 30 were engaged in the trade. A recent forest-congress made the following suggestions :

1. The prohibition of the import of Manchurian wood for the railways.
2. A duty on all Manchurian wood.
3. A minimum tariff for the export of wood by the Ussuri Railway.
4. Better naming and classification of timber.
5. Free choice of labour. This means in practice the right to employ Chinese.

The enforcement of the first recommendation would do

little good because the railways use mainly hard wood—oak, larch, and birch—whereas the Priamur produces chiefly soft wood, such as pine, fir, lime.

Classification of timber.—The *kedr* or white cedar is the only tree at present in such demand that it is cut in large quantities. The best forests in the Maritime Province are found near the sources of the Iman and Khor. The grain is much finer than in Manchuria. The average of square logs is 19 inches, the maximum 28 inches. In Vladivostok it fetches from 7½*d.* to 8*d.* per cubic foot, but from 60 to 70 per cent. of this represents railway freight. In the harbours of the Maritime Province the cost is from 5¾*d.* to 6¼*d.* alongside vessel. Most of the best goes to Great Britain.

Larch suitable for use under water or for telegraph-posts is to be had throughout the northern regions of the Priamur. It costs about 5*d.* to 5½*d.* alongside vessel. Larch, fir and spruce, with some birch and aspen, make up the northern forests. The price may be given as 4¼*d.* to 4¾*d.* on board. Oak and ash of moderate quality are found in many places, but are used only for firewood. The oak is said to be rich in tannin. The oak in the interior of the Maritime Province is of better quality.

The pitch pine (*Picea ayanensis*) is fairly close-grained, white, and very light, and is exported in large quantities to Australia. Excellent yellow pine is found in the very large forest between Blagovyeshchensk and Chita, and also along the Zeya and the Bureya. It is at present quite untouched, but the cost of cutting and rafting makes it doubtful whether it could be delivered at Nikolaevsk for 5½*d.* per cubic foot, as suggested.

In any case only the higher grades of wood, from 18 to 20 per cent. of the total output, are suitable for export to Europe. The Government takes most of the remainder.

Home Consumption.—The local needs are considerable. The railway and the steamers take more and more every year. A large quantity of timber is annually floated down the Zeya and the Bureya, especially to Zeya-Pristan, Blagovyeshchensk, and other stations, for use in the mining camps or on the railway. Moreover, the capture of the Amur fishing industry by the Russians from the Japanese has greatly increased the demand for wood for packing purposes, and a certain amount is needed for match-making. There is no attempt at reafforestation.

Exports.—The export trade is growing steadily and the tar industry developing. The total amount of timber exported from the Priamur in 1910 was 1,617,650 cubic feet, in 1912, 2,272,570 cubic feet. Vladivostok is still the chief timber port, though the fact that the timber must be brought there by rail is against it. In 1912 more than half the timber was exported from Vladivostok, 687,548 cubic feet being sent to Great Britain, 316,624 cubic feet to Japan, 37,070 cubic feet to China, 51,997 cubic feet to Korea, and only 1,545 cubic feet to Russia. By far the greater part of the wood exported was *kedr*, except to Japan, which took 307,204 cubic feet of aspen for matches. Vladivostok is also the natural outlet for the excellent Manchurian timber, and a considerable portion of the exports come from over the border. The cost of sending the timber by rail and the absence of adequate docks are serious drawbacks to Vladivostok. But preparations are being made for building a new timber-port, where four vessels can load at once.

Imperatorskaya Bay, whence the Oriental Timber Company sends over a million cubic feet of timber (25 per cent. larch, 75 per cent. fir) annually to Australia, is the only other timber-port of note. Exports to Australia are likely to increase steadily, for Australia is looking to Siberia to supply her with the soft woods which her own forests do not produce in sufficient quantities. This company also sends planks and furniture wood to the British Isles, ash to Japan, and logs and boards to Denmark. In 1911, however, 110,000 cubic feet of white pine were exported from Ternei Bay to Australia, and 374,000 cubic feet of pit props were dispatched from Olgi Bay. In the same year timber was exported from Posiet Bay for the first time, to the amount of about 100,000 cubic feet. Tyutikha Bay also plays a small part in the timber-trade.

The Governor-General Gondatti was fully alive to the importance of the industry, and on the completion of the Amur Railway he proposed to agitate for a branch line from the Ussuri Railway to Imperatorskaya Bay and Olgi Bay.

Aspen, which is found everywhere, is sent to Japan for match-making, but there is no reason why the industry should not be established in Siberia. The oak of good quality, but small dimensions, which grows in some parts of the interior of the Maritime Province in small quantities, is beginning to be exported.

With better management the Priamur forests should be able to supply China with much of the timber she now imports from Japan and the United States.

Sawmills

The annual turnover of the sawmills in Siberia is about £300,000. The principal ones in western Siberia are at Tobolsk Tyumen, Omsk, Novo-Nikolaevsk, and Tomsk, with others serving a smaller radius at Barnaul and Biisk. The great sawmills at Novo-Nikolaevsk deal principally with wood from the neighbourhood of the Ob, between Barnaul and the railway. The wood is cut into logs of 21 feet. The best wood has a diameter of from 12 to 16 in.; trunks are found with a diameter of 35 in., but in that case the heart is no longer good. In big sawmills that work day and night as many as 20,000 trunks are sawn through in the year at each frame. The railway greatly stimulated the amount of sawmill work, for much of the wood was used on it, and it provided means of transport. Sawmills are usually lit with electric light.

In the town of Irkutsk the sawmill industry is considerably developed. The wood comes from Lake Baikal and the valleys of the Angara and Irkut. Logs are sent even to Transbaikal.

Blagovyeshchensk, where there is a prosperous sawmill industry, is the centre of the timber trade on the Amur. The Government has already established three mills of Swedish type on the north bank of the Amur, and has taken over another at Sviyagino on the Ussuri Railway. It also proposes to build a large mill of the latest American pattern, but it has not yet selected the site. At Alexandrovsk in Sakhalin there is a government sawmill supplying local demands. In 1912 the sawmills in the Amur and Maritime Provinces and Sakhalin numbered 62, with an annual output of 2,050,000 logs.

Wood Industries

The manufacture of veneer and three-ply in the neighbourhood of Vladivostok is steadily increasing. Several factories for the making of barrels for the Amur fishing industry have been established in and round Vladivostok.

At Spasskaya (Yevgenevka), on the Ussuri Railway, a factory exists for the chemical treatment of wood and the production of turpentine, tar, wood alcohol, vinegar, resin, and potash.

CHAPTER XII

MINERAL RESOURCES

Iron—Copper—Gold—Silver—Zinc and Lead—Platinum—Asbestos—Graphite—Mica—Petroleum—Other Metallic Ores—Coal—Salt—Precious Stones and Building Materials.

THE mineral resources of Siberia, especially of the east, are very little known. The merest beginning has been made with their development. But they will undoubtedly take the first place in attracting much-needed capital to the country in the near future, especially to the remoter regions.

IRON

The iron deposits in the Urals are second only to those of south Russia in the part they play in the iron industry of the empire. The output of pig-iron from the Urals in 1913 was 896,817 tons. There were 75 iron works, including 13 belonging to the Government, at that time active in the province. The output has fallen off somewhat during the war.

In western Siberia the railway has proved fatal to such attempts as have been made to establish an iron industry. The demand is not sufficient to support a large foundry. The old mines away from the railway cannot now compete with the Ural foundries. The Bogoslovski works, for instance, can send their goods into the heart of the Ob basin on their own steamers by the Tavda and the Irtish. The Tomsk region perhaps offers the best prospects for the establishment of a successful iron industry now that the Altai Railway is open. On the Telbes, a tributary of the Kondoma, near Kuznetsk, are rich deposits of magnetic ore within

20 miles of beds of good coking coal. The deposits at Gurevskoe, some 150 miles to the north of these, only contain inferior ore and do not promise a prosperous future to the small foundry that works them. All the Altai beds are Crown property. There is also a small factory at Abakanskoe in the Minusinsk region, where there are plentiful deposits of ore, but its output is diminishing. Some twenty iron beds have been located in the Kirghiz steppes, more especially near Karkaralinsk. They have not yet been properly examined and are too remote for profitable working, but in some of them the percentage of ore is high and there is coal near. Iron deposits also exist near Tyumen, from which the owner annually extracts his statutary 170 tons by the most primitive methods in order not to forfeit his concession.

In eastern Siberia even the Nikolaevsk foundry, at one time the largest in the country, situated on the River Oka, 30 miles from Bratski-Ostrog, in the neighbourhood of excellent ore, has been obliged to close down. Iron is found in several parts of the valleys of the Yenisei and its tributaries, such as the Abakan and the Angara; in the valleys of the upper Lena and the Kirenga; near Misovsk on Lake Baikal; and on the Tsagan-Khuntei Range, west of the Khilok valley, near where the range is crossed by the post-road. In the Nerchinsk district several deposits are known, but they have been little investigated. It is so plentiful in the southern regions, which are the centres of the gold and silver mining, that the whole range there has been called iron. The Baleginsk deposit of magnetic ore supplies the Petrovsk iron works which are on the railway. They were built to supply the needs of the Nerchinsk Crown lands and have never been very productive. The machinery is quite obsolete. Hitherto only red oxides have been treated. The deposits are extensive and there are said to be large supplies of magnetic ore quite untouched in the neighbourhood.

Iron is plentiful round Yakutsk. It also exists in the valley of the Amgun, near the mouth of the Amur, where it should be capable of profitable working, round Olgi and

Vladimir Bays, as well as in the peninsula of Kamchatka and near Dué in Sakhalin. A rich deposit has recently been discovered near the junction of the Samara and the Amur. The irregularities of the compass between Plastun Bay and Cape Povorotni point to the existence of extensive iron-beds in the Maritime Province.

The deposits near Vladimir and Olgi Bays are commercially the most promising, though the nearest fuel, the timber on the Sikhota-Alin, is 13 miles away and Olgi Bay is not too good a harbour. Hitherto they have been almost unexploited, but now that the export of ore is no longer prohibited, a profitable market lies ready to hand in Japan. A blast furnace is to be built at Mramorni Point, near Olgi Bay. Eastern Siberia has till now been largely supplied with pig-iron from Germany.

Copper

Copper has been worked in Siberia from time immemorial. Prehistoric or *Chud* excavations, as they are called, are frequently found in the west. They often gave the first indication of the presence of the metal to the early prospectors.

The output of copper in the Russian Empire rose from under 10,000 tons in 1906 to 34,300 in 1913 and there is no reason why the country should not supply its own needs at an early date. The industry is protected by a duty on imported copper and the price of copper in Russia is high in consequence. The rapid growth of the output dates from 1907, when the Myed (copper) Syndicate, which combined the companies and regulates the trade, came into existence.

The Urals hold the first place; then come the Kirghiz steppe and the Caucasus. These three regions produce over 90 per cent. of Russian copper. In 1913 the Urals produced some 16,000 tons, the Caucasus 9,900, and Siberia 5,600. British capital has played an important part in the development of copper-mining. The Kishtim Corporation, a British concern, produces nearly half the copper in the Urals and more than one-fifth of the total Russian output. It is the only company in Russia which smelts its own ore and refines it at its own works. Next to the Kishtim rank the Bogoslovski works, which are the oldest in the Urals, having been founded by Demidov in the eighteenth century. There are other works in the Tagilski, Iset, and Sisertsk districts. Verkhoture is a considerable smelting centre.

In the last few years diamond drilling, carried on by British firms, has revealed large bodies of copper-bearing pyrites. The Russians are adopting this method. Experiments are being carried out at one of the Ural works for saving the sulphuric gases given off during smelting.

In the Kirghiz steppe at least 200 outcrops of copper have been located, mostly in the Karkaralinsk district. The deposits are rich and coal is abundant. The absence of railways makes it impossible to work anything but rich seams at a profit. The Spassky Company, which is now in British hands, and has recently acquired control of the Atbasar works, is by far the largest and most successful in this region, in spite of the difficulties of the conditions of working. In 1914 the output was 4,683 tons; in 1915, 3,450 tons. The construction of the Southern Siberian Railway should bring about a great improvement. There are also copper deposits in the Semipalatinsk region.

In eastern Siberia copper has been found in the Minusinsk region, near Verkhne-Udinsk, and in the Argun and Onon basins, where, however, the veins are poor. In the Maritime Province there are deposits near Vladivostok which produced 13 cwt. in 1911, as well as near Konstantinovskaya on the Suifun, and round Dzhigit Bay, where they contain as much as 80 per cent. of ore. Copper-ore has also been discovered near the mouth of the Kolima, at the confluence of the Big and the Lena, in the peninsula of Kamchatka, and elsewhere.

Gold

Gold-mining is steadily declining in importance in western Siberia, but in the east it is the one productive industry. It has done more to open up remote districts than all the emigration agencies. Nothing but a gold-rush could bring about the sudden rise of towns like Bodaibo or Zeya-Pristan.

Extent of deposits.—The extent of the gold deposits in Siberia, but especially in eastern Siberia, is very imperfectly known. Gold is found in the alluvial deposits of many of the rivers, but only occasionally in sufficient quantities to repay working by old-fashioned methods. Areas suitable for dredging extend for thousands of miles in the eastern provinces; and if modern dredgers were employed, the results would be surprising. At present the chief centres of the industry are the Urals, the lands which belonged to the Imperial Cabinet in the Barguzin and Nerchinsk districts, the Zeya and Bureya basins and their neighbourhood, the Amgun system and the mines near Lake Chlya, and lastly the Olekminsk and Vitim goldfields on the Lena. Of these the Vitim mines are by far the most important, producing a quarter of the total gold output of the Russian Empire. The Lena drainage area is said to contain the richest alluvial gold district in the world.

The principal areas of gold-bearing rocks in Siberia have been estimated as follows:—

	sq. miles.
Urals	60,000
Yenisei and Altai	280,000
Transbaikal	20,000
Lena, Vitim, Vilyui	160,000
Amur and Okhotsk	360,000
	880,000

This may be compared with the total area of gold-bearing rocks in the United States, including Alaska, of 265,000 square miles.

Output.—Between 1908–13 the average annual output of gold in Siberia was one and a half million ounces troy, with a marked tendency to decline, especially in western Siberia. The official estimates are not to be trusted. The amounts received at the Government laboratories where the gold is smelted, especially in the east, are often double those recorded and a good deal of the metal never reaches the laboratories at all. The official estimates are as follows :

	1910.	1911.
	lb. troy.	*lb. troy.*
Blagovyeshchensk	21,548·7	19,012·5
Bodaibo	39,153·6	36,251·1
Nikolaevsk	4,140·9	5,098·5
Zeya-Pristan	—	2,615·4
Krasnoyarsk	2,889·0	1,525·4
Private laboratories	23,070·6	14,502·6

The figures for the Yekaterinburg, Tomsk, and Irkutsk laboratories for these years are not to hand.

In 1913 the eight Government laboratories of Siberia smelted 106,901 lb. troy of gold ore compared with 104,155 lb. in 1912. The increase was due entirely to the settling of the Lena strike. Bodaibo produced 7,200 lb. more than in the previous year (1912), when the output was 24,704 lb. troy.

Conditions of working.—With the exception of Alaska, Siberia is the least favourably situated of the world's gold centres. Owing to its climate, work is only possible in summer. Most of the gold is found in places where the ground never really thaws. If the summer is dry, there is not enough water for washing ; if it is cold and wet, the ground does not thaw and floods are troublesome. Population is everywhere scanty and roads almost non-existent. The cost of bringing machinery, especially in summer, from the railway to the diggings is almost prohibitive, while the prices of the most ordinary necessities of life are doubled and trebled.

Concessions.—No foreigner may possess freehold property in Siberia, and a special tax of 8 per cent. is levied on land or mining claims leased to foreigners, who may not prospect for gold within 100 versts (66 miles) of the sea-coast of the Primorsk Province or of the shores of the Sea of Okhotsk, upon Sakhalin or the adjacent islands, or within certain districts of the Yeniseisk Province bordering on the Chinese frontier. In this region foreigners have considerably less freedom than Russians, as well as in the Cabinet lands in the Altai and Nerchinsk districts. But even here a foreigner may receive special permits for prospecting. A prospector may stake out as many claims as he likes. One man has been known to possess over 750. But he must take measures for working them within a certain time.

There is a large traffic in worthless claims. The local officials, to whom commissions must be paid, are often untrustworthy. Much foreign capital has already been lost in Siberian gold-mines and investors are chary of risking more.

The principal British gold mining companies in Siberia are the Lena Goldfields, the Orsk Goldfields, and the Siberian Proprietary Mines. The Lena Goldfields held until 1915 over 50 per cent. of the shares of the Lenskoie Gold Mining Company of Russia. They have now sold the greater part of their holdings.

State of the Industry.—Except in the Urals and to some extent in the west, the gold worked in Siberia is still almost entirely alluvial. Though rich quartz deposits are known to exist, quartz mining could only be profitably carried on at a few places in the west. But when the country is better developed there will be a great future for such mining.

As the more accessible placers become exhausted, the general tendency is for large concerns to supersede small ones. More scientific methods are now essential to success. Dredgers alone can work profitably and they are not easy to procure. The Russians have little practical knowledge of mining and can only make the richest claims pay. The incapacity of the Russian engineers and workmen, many of whom have never seen a dredger, accounts for the frequent breakdowns and poor results. When dredging is regularly established and thoroughly understood, many of the abandoned dumps will be profitably worked. The latest methods of working hydraulic sluices are unknown in Siberia, where the climate militates against their introduction.

At present the Siberian gold industry is going through an unsatisfactory period of transition. The old methods are no longer adequate and the new are only beginning to be introduced. In 1912, 57 mines out of a total of 129, belonging to 48 different companies, were working in the Amur district. A committee of investigation was appointed and various suggestions were made such as the regulation of the labour-supply, the diminution of taxation and greater facilities for credit, as well as the throwing open of the 100 verst reserve along the coast. Owing to its accessibility, this region would rapidly attract prospectors. Between 1889 and 1909 mining machinery was imported duty free and the abolition of this privilege was undoubtedly a blow to the gold industry. In recent years the Government proposed to assist it by voting large sums for prospecting and for building a number of dredgers.

But the success of companies with large capital and good machinery, such as the Lenskoie and the Orsk, shows the lines along which the gold industry will ultimately develop. It is noteworthy that American engineers are more successful than British, who have gained their experience in South Africa and know nothing of the conditions of Russian mining. British companies are now employing them.

Labour.—The labour-supply lies at the root of the difficulties. Railway building attracted a number of miners and the Government is doing its best to exclude the Chinese wherever possible. A man who could prove that he had worked on the gold-fields was exempted from military service in the present war. The life of the miner is hard in the extreme. He is as often as not obliged to work in water. Except in a few of the largest concerns, the workpeople are not well treated. They are provided with firing and lodging, but their quarters are often badly overcrowded and invariably filthy. The managers are tyrannical. The men are compelled to buy at the company's stores. If they complain or threaten to go elsewhere, they are liable to be dismissed. It is true that a law of 1902 declares that the labourer is to have suitable quarters and to be supplied with food and clothing at special rates, which vary in different districts. But the scandals revealed by the Lenskoie strike show how easily these regulations can be evaded in the remoter districts. The mine stores are often more profitable to the owners than the mines themselves. When the lot of the Siberian miner is contrasted with that of the well-paid, well-fed miners of the Alaska gold-fields, it is not surprising that the work does not attract the most desirable elements. The miners are continually running away. The building of the Amur Railway has greatly thinned their numbers. The sale of alcohol is absolutely prohibited, though a small quantity is distributed free to the men. But there is a large illicit traffic in liquor, which the mining police are unable to check.

All gold visible to the naked eye is the property of the workman who discovers it and he receives 75 per cent. of its value. This unprofitable arrangement has been adopted to discourage theft. In their spare time the men may wash for themselves, on condition that they sell their gold to the companies at a fixed price. Indeed, many companies habitually leave all but the richest ground to be worked by private individuals, generally Chinese or Koreans in the eastern provinces, who pay a rent of gold to the owners and sell them whatever additional gold they may procure at fixed rates. This arrangement is very profitable to the companies. But there is, of course, a large illicit gold trade for which the sellers of contraband alcohol are largely responsible. It is smuggled over the border by Chinese, who also smuggle in Chinese spirits. At a recent mining congress it was said that in the Transbaikal, where most of the gold-mining is in the hands of Chinese and Koreans, they carry away seven times as much as they declare. This may well be an exaggeration, but it throws much light on the prevailing state of affairs.

Urals and Western Siberia.—Gold-mining is on the decline in these districts. As everywhere else in Siberia, dredgers are becoming more and more necessary for successful placer-mining, but the Ural fields are held to be of low grade. In the more accessible districts quartz-mining is steadily superseding alluvial mining. In the Urals gold procured by quartz-crushing and saved by chemical processes already exceeds the gold obtained by washing. In 1913 the Urals produced 110,000 oz. troy of gold, worth £399,208. The Orenburg and Yekaterinburg districts are the richest. The mines in the Miyas district still hold the first place in the Urals. The Sissert Company and the Kishtim Corporation both obtain gold as a by-product in their copper-mines.

In the Semipalatinsk Government there is considerable activity in gold-mining, but here too placer-mining is giving way to quartz-mining. The quartz veins are especially rich in the neighbourhood of Ust-Kamenogorsk, less so round Lake Zaisan. Near Ust-Kamenogorsk the quartz yields nearly 5 to 15 dwt. to the ton, but the gravel produces barely 5 grains to the ton. Hence, in spite of the cheapness of labour, placer-mining can hardly pay. The richest mines lie 40–46 miles south-west of Ust-Kamenogorsk. As this district is within the zone closed to foreigners, they can only prospect there by special permission and the native owners are fully aware of the value of their holdings. There is a good steppe road from the mines to Ust-Kamenogorsk.

The gold-industry in the Tomsk district is in full decline. Even the introduction of hydraulic machinery has failed to revive placer-mining in the Altai. A British company is experimenting in mining for gold in the old cabinet silver-mines in the neighbourhood of Zmyeinogorsk in the same region, and if it succeeds it will certainly extend its operations. The only prosperous gold-mining region is that of the Mariinsk taiga. The first dredger in western Siberia was established here. The output of placer-mining is falling off, but rich veins have been discovered in the quartz of the Berekul mine. These are being worked, and, according to official returns, they yielded in 1904 about 1 oz. troy to the ton. The veins had not then been thoroughly examined. The mines are well situated, being 45 miles from the railway and 65 from Mariinsk.

Yeniseisk Province.—In the south-west corner, near the River Abakan, are several mines. Of these the Bogom-Darovanni, fitted with thoroughly modern machinery, is very rich. It is one of the few profitable reef gold mines in Siberia and it produces some 17,100 oz. troy of gold annually, a quarter of the amount being obtained by the cyanide process.

The introduction of dredgers has revived the gold-industry in the Yenisei basin, which was the oldest in Siberia. The southern portion of the district lies between the Pit and the Angara Rivers ; the northern is to the east of the Yenisei, in the upper basins of the Teya and the Kalami, tributaries of the Stony Tunguska. The pay-streak is from 2 ft. 4 in. to 8 ft., the overburden from 2 ft. 4 in. to 24 ft. The district is remote, but the mine owners refuse to combine to build a road through the taiga, though the cost of transport is at present very heavy. Many of the mines in the south are connected by telephone.

Irkutsk Province.—There are no mines of importânce south of the railway. The Olekma-Vitim system embraces all but a negligible quantity of the mining.

The Bodaibo district is the most important gold-mining centre in the Russian empire. The powerful Lenskoie Company, formerly controlled by British interests, has a virtual monopoly. It produces some 13 tons of gold a year, which is a quarter the annual output of the whole empire. The gold is alluvial. The pay-gravel lies from 50 to 150 ft. under the surface and the streaks are from 4 ft. 8 in. to 9 ft. 4 in. thick. The over-burden is peat. The placers can only be worked with considerable capital. Wood is scarce in the neighbourhood and the subsoil water requires careful regulation. The yield of gold is from 82 to 205 grains to the ton of gravel. About 4,000 men are employed here and up-to-date machinery is used. The transport difficulty adds greatly to the working expenses, but there is now a light railway 15 miles long from Bodaibo to the Vitim. It costs 1*s*. to bring 12 lb. of goods from Irkutsk to Bodaibo alone. Hence the mining companies are among the chief advocates of the building of a railway to the Lena.

Most of the tributaries of the Vitim are thought to contain gold, but they have been very imperfectly explored.

The Olekminsk mines were formerly the richest in Siberia, but since the best placers have become exhausted they have been thrown into the shade by the mines on the Vitim. The River Bolshoi-Patom, one of the richest centres on the Lena, is said to have yielded 14,000 oz. troy of gold in 1911 and the whole of its banks have been staked out in claims. The building of the Lena Railway would undoubtedly initiate a new period of prosperity for this region.

Gold has been found on the upper reaches of the Vilyui and its tributaries. The gold-yield at Chodinski, some 80 or 90 miles above Krestyatskaya, is said to be large. Gold also exists on the upper reaches of the Nai, a tributary of the Aldan.

The government assaying and gold-smelting laboratory is at Bodaibo.

Transbaikal.—There are two important centres, Barguzin and Nerchinsk. The Barguzin goldfields lie in the Barguzin valley and near the sources of the Vitim, but of recent years the output shows a considerable falling off. Though the gold-yield is sometimes 41 to 82 grains to the ton, the veins are very small and there is a quantity of silver mixed with the gold. Belgian engineers are trying the experiment of thawing the ground here by a process of steam heating.

Gold is found almost everywhere in the crown lands in the Nerchinsk district, except in the south-east and north-east. The mines possess the same characteristics as those round Barguzin. British companies have been experimenting lately with a view to leasing. Attempts to introduce the most modern machinery have not been very successful owing to climatic difficulties. The output of the Transbaikal was about 171,000 oz. troy in 1909.

Amur Province.—For mining purposes the Amur Province is divided into two districts, the Bureya and the Amur. At present the former is the more important. In 1911 these districts produced £443,830 of gold, of which five-ninths came from the Bureya. The Amur goldfields are of far greater extent than those of the Lena. Alluvial gold has been found throughout the basins of the Zeya and its tributaries, and in the Bureya basin, especially upon its tributary the Niman. New deposits were discovered during the building of the western section of the Amur Railway.

Most of the Amur gold is so fine, yielding from 0·04867 to 0·13114 oz. troy to the ton of gravel, that only the best machinery could make mining profitable. A quarter of the diggings worked yield over 40 grains of gold to the ton, but the pay-gravel is often less than 2 ft. 4 in. deep, sometimes even less than 7 inches. When the yield is less than 40 grains to the ton, and the pay-strata are less than 2 ft. 4 in. thick, profit is very doubtful. These strata often lie under an over-burden of 9 ft. or more, and in some places, notably on the Niman, the mining is underground.

Holdings are often very large in the Amur Province. The Upper Amur Company owns goldfields in the Rivers Zeya, Zhalinda, Gilui, and Aldan. At least a third of those worked are sub-leased. Little machinery is used. The Orsk Company has claims on the lower Amur.

Zeya-Pristan, with a government gold laboratory, is the flourishing capital of the Zeya gold industry. Blagovyesh-chensk, at the mouth of the Zeya, also possesses a laboratory.

The opening of the Amur Railway should greatly benefit gold-mining in this region.

Maritime Province and Kamchatka.—The Primorsk region is the wealthiest. In the Amgun basin, near Kerbinski, the mining is all surface mining. The pay-gravel is from 4 ft. 8 in. to 7 ft. thick, the over-burden 3 ft. 6 in. to 14 ft. Owing to the nearness of the coast and the scarcity of labour machinery is much more used. But difficulties of communication and the dearness of food make a high yield of gold necessary for success. The Amgun mines were long worked at a loss. The Orsk mines, owned by a British company, situated on Lake

Chlya, near Nikolaevsk, are now the most successful in the neighbourhood. Two powerful electric bucket-dredgers are working, and the profits, even during the war, have steadily increased. In 1912, 21 placers were being worked here, and the mines were responsible for a quarter of the £216,604 of gold produced by the province.

The Ussuri district comprises the southern part of the province, and here Chinese and Koreans have long ago exhausted the more accessible deposits. In 1911, £6,818 of gold was produced, in 1912 only £1,916. Recently, however, placers have been discovered on the Iman, which are estimated to yield £1 per ton of gravel. A small quartz vein on Askold Island near Vladivostok was profitably worked for some years.

The official gold-smelting laboratory for the Maritime Province is at Nikolaevsk.

Gold is known to exist at many places on the Sea of Okhotsk, and is said to exist in the centre of Sakhalin. A gold-bearing belt is believed to extend for some 120 miles along the Okhotsk coast between the Uda and Ayan. Rich deposits are reported in the Anadir region, in the Chukchee Peninsula near Cape Dezhneva, and on the River Volshaya in the Anadir region. The Volshaya mines have remained idle since 1907, but the deposits are said to contain about 240 grains of ore to the ton of gravel, and might therefore be profitably worked, in spite of the remoteness of the region and the absence of wood. A large expedition was sent out in 1914 with a view to reopening them.

Silver

The output of silver in Russia reached its zenith in 1887, with 33,800 lb. troy. In 1910 the output was 19,476 lb.: Urals, 13,356 lb.; Caucasus, 5,508; Altai, 612. The fall in the value of silver, the discovery of rich goldfields, and the labour difficulties, account for the diminution. In the Urals silver is produced chiefly as a by-product, notably in the Kishtim, Blagodat, and Verkhne-Iset mines. There are many other deposits. Over 3,000 deposits are known to exist in the Altai, of which only some 30 have been worked. The Ziryanovskoe mines are the richest, but even they are almost shut down. There are others in the neighbourhood of Zmyeinogorsk. The silver-works in the Kirghiz steppes are very primitive. In the south-eastern portion of the Nerchinsk Crown lands 500 deposits of silver-lead ore are known to exist. The richest are the Kadainskoe, on a tributary of the upper Argun. Silver is also found in the zinc mines of Tyutikha Bay (see below).

Zinc and Lead

The zinc and silver-lead mines near Tyutikha Bay in the Priamur have been most successfully worked. In 1911 the output was 24,030 tons of zinc, 4,451 tons of silver-lead, and 72 tons of copper ore. In 1912, 25,000 tons of zinc were sent to Europe, mostly to Antwerp, to be smelted. But the late Governor-General, Gondatti, insisted that henceforth the smelting should be done on the spot, and a smelting furnace was consequently built. The Tyutikha mines lie 24½ miles from Tyutikha Bay, with which they are connected by a railway of 60 centimetres gauge. The ore contains nearly 50 per cent. zinc, little silver, and about 5 per cent. copper. Other claims are being taken up in the neighbourhood. The Tyutikha company has just located new deposits near Imperatorskaya Bay.

Zinc and lead ores are being mined in the Altai by the Irtish Corporation. From Riderski mine the ore is taken by a 3-ft. railway, 70 miles long, to Ust-Kamenogorsk, whence it is shipped by the Irtish to the smelting works at the Ekibas-tuse coalfield near Pavlodar.

Zinc is also said to exist in the Yakutsk Province. Lead-ore has been found at Ust-Orlinskaya on the Lena, where the content is said to be 81·75 per cent. of pure lead. Silver-bearing lead also exists in the Altai and in the Nerchinsk district, but with the present means of communication it is not likely to be exploited.

In 1915 eastern Siberia produced about 48,300 tons of lead, as compared with 30,000 tons in 1913.

Platinum

Ninety per cent. of the world's platinum comes from the Urals. But in spite of the high price the output has fallen from 99,820 oz. troy in 1912 to 79,000 in 1916. The richest placers are rapidly becoming exhausted, and the others can only be worked at a profit with dredgers. More than half the output comes from the Verkhoture region, notably from Tara River on the Tagilski estate. Platinum is found in some quantities on the Sosva and Lozva and other rivers. Traces of it occur in the Tomsk and Mariinsk districts, notably in the gold mines of the Chumish basin, in the Yeniseisk goldfields round the Pitski Mountains, near the Vitim goldfields, and on the Uni Bolski in the Amur Province. The natives in the Aldan valley are said to use it for bullets. Hitherto platinum has been sent abroad to be refined, but in 1915 an export tax of 15 per cent. *ad valorem* was placed upon unrefined platinum with a view to encouraging the erection of refineries at home.

Asbestos

Asbestos is mined almost exclusively near Yekaterinburg, and before the war was virtually all exported to Germany and the United Kingdom via Riga. The output in 1913 was 16,661 tons. In the Irkutsk Province asbestos is only worked in the Angara district. The quality is good. On the Mongol-Dabanski gold-placers, which belong to the Crown, but are now worked out, there are very rich asbestos and mica mines. They are on a tributary of the Didi, which is a tributary of the Oka, and are 75 miles from Ziminskoe. In Yeniseisk there are asbestos mines on the left of the River Kamishta, a tributary of the Abakan. The asbestos lies in dolomite veins up to 4½ ft. in thickness, but only one-seventh is of commercial value. It is found also on the River Karagan in the same government, in the Altai; 150 miles south of Biisk, on the Katun; in Transbaikal near Shilkinski; and in the neighbourhood of the Nerchinsk tin-mines.

Graphite

Hard and clean graphite has been found in considerable quantity near Turukhansk on the Yenisei, and 130 miles west of Irkutsk, on the Mongolian border, as well as in the Kirghiz steppe. Graphite of excellent quality comes from the neighbourhood of Cape Dezhneva. It also exists on the Lower Tunguska, near the River Bukhalova and near Souznaya on the Amur. But nowhere is it exploited to any extent.

Mica

Mica is found in the Krasnoyarsk region, but has not been systematically worked, at Kandakova on the River Tasyeeva, a tributary of the Angara, as well as on the River Kan near the mouth of the River Varga. It is also reported in the Nizhne-Udinsk district, and on the Mama, a tributary of the Vitim. On the southern shore of Lake Baikal the quality of the mica found is good and the pieces usually large.

Petroleum

A belt of large petroleum-bearing rocks about two-thirds of a mile wide stretches along the shores of Lake Baikal as far north as Barguzinski Bay. In Sakhalin naphtha springs exist near Niski and Nabilski Bay, close to harbours accessible to sea-going vessels. There are lakes of petroleum near the Nutovo River, and petroleum which flows of itself has been discovered at a depth of 1,000 ft. A larger oil-sand should be reached at 2,000 ft. The oil is said to be without benzine, and it can therefore be used for fuel immediately. Expert opinion compares it favourably with that of Baku.

Other Metallic Ores

Tin of good quality has been found near Nizhne-Sharonai, Olovyannaya, and elsewhere between the Onon and the Ingoda, but it is little worked. There is every sign that this region is a genuine tin district. The Government recently was said to be installing here the first tin-smelting works in Siberia.

Antimony.—The important antimony deposits at the Akhatolskaya mines in the Urals are being carefully investigated. Antimony also exists in the Urals near the Verkhne-Neivinsk works and in the silver mines 10 miles north of Blagodat. It is found in several mines in the Yeniseisk Government, and in more than one river in the Minusinsk district. In the Transbaikal a spring near the Kadainskoe silver mines is known as the antimony spring. Antimony deposits stretch along the mountain top 7 miles from Zabilovo in the Amur Province. The vein reaches a thickness of 3½ ft.

Mercury.—Quicksilver has been discovered in the Verkhne-Iset district in the Urals; the veins seem numerous and rich, but have not been fully investigated. Deposits of cinnabar are said to exist near Lake Ayamskoe. Cinnabar also occurs in the Bogoslovski district; at Ildekanski, in the valley of the Urov, a tributary of the Argun, where the vein is rarely more than 2 inches thick; in the Amga basin and in Kamchatka. But it has not been worked in these regions.

Radium has been found on the Ayakhta, a tributary of the Pit, which flows into the Yenisei. The Kamchatka province is believed to have deposits of *iridium*, *palladium*, and *osmium*. *Thorianite* has been discovered in the black slimes of the placers on the River Boshagoch in the Nerchinsk region. Rich deposits of *wolfram* exist in the Urals and near Klyuchevskaya on the Ingoda, and near the Onon in the Transbaikal.

Manganese is found at Nizhne-Tagilski and elsewhere in the Urals, and is said to exist in the valley of the Angara. *Molybdenum* is found in the Kirghiz steppe and Transbaikal.

Osmiridium occurs in the Kishtim mines in the Urals, in the Nizhne-Udinsk district, in the Dzhila river system in the Transbaikal, and in the gold-placers of the Troitskosavsk district. It is nowhere regularly worked.

COAL

There are large deposits of coal in Siberia, but most of it is only of small importance. Many of the deposits, however, are not yet worked, and it is probable that some of the coal in eastern Siberia will prove to be of good quality. In many coalfields the want of a local market prevents extensive exploitation. While the population is scattered, as it is in most parts of Siberia, and while timber is abundant, coal cannot be produced cheaply enough to be used as fuel except in the immediate neighbourhood of the beds. The railways are the great consumers, and are steadily increasing their demands. Manufacturing industry is still too little developed in Siberia to require great quantities of coal, and export is economically practical be only from the coalbeds of the far east.

Urals.—Coal occurs on both flanks of the Urals. All these deposits are properly speaking in European Russia. The beds on the western side have been worked for many years, but those on the eastern side are only beginning to be developed. The wholesale destruction of forests, and consequent increased demand for, and enhanced price of coal makes it probable that these deposits will play an important part in the industrial activities of the Ural region. Unfortunately, however, this coal will not coke, and is therefore useless for metallurgical works. Hence coal is imported from the Donetz region of southern Russia. In 1914 the production of coal in the Urals was 1,170,412 tons, and was increasing annually. On the other hand, in some factories peat is being used as fuel, and machinery for compressing peat has already been installed in places.

On the lower Ob in the neighbourhood of Berezov is an extensive coalfield, but up to the present this has not been worked.

Kirghiz Steppe.—East and west of the Pavlodar-Karkaralinsk road are a number of coal outcrops which point to the existence of large deposits apparently scattered in isolated basins of small size. Only the Karagandinsk mine, 134 miles north-west of Karkaralinsk, which supplies the Spasski copper works, at present possesses commercial importance. The coal has moderate coking qualities, but gives nearly 40 per cent. ash. The seams are from 3 to 20 ft. thick. The Ekibas-tuse mine is being worked on a large scale, and a railway has been built to the Irtish, by which zinc and lead ores are brought for smelting from Riderski in the Altai. The coal is coking coal, and there are two seams of variable thickness which have been traced for over 4 miles at a depth of about 400 ft. The lower seam has little ash. The output in 1917 was 80,000 tons. The completion of the Southern Siberian Railway will give increased importance to these mines. They are owned by a British company, the Irtish Corporation.

Kuznetsk.—The Kuznetsk beds, in the Altai region, extend from Sudzhenka on the railway to 40 miles south of Kuznetsk. On the west they reach the Ob in places, on the east the slopes of the Ala-tau Range. The total area is estimated at about 5,000 sq. miles. The quality varies considerably in different parts of the beds. Floods and the irregularity of the seams cause trouble at the State mine at Anzherski. The seams, ten in number, are from 3 to 45 ft. thick. The Sudzhenka coal, which is coking or semi-anthracite, is taken almost entirely by the railway. Small quantities only are sent down the Tom for local use in Tomsk from the Kolchugino mine, which supplies the inconsiderable needs of the ironworks in the neighbourhood. The shaft here is 25 ft. deep. A French company working this coalfield failed, and there is now no mining. The Yurga-Kolchugino Railway, now extended to Kuznetsk (see Chapter XVII), should at last make it possible to exploit these valuable coal beds.

Cheremkhovskoe.—Important beds of coal lie round Cheremkhovskoe, 70 miles west of Irkutsk. They produce nearly 5,000,000 tons annually. The coal is mainly lignite, and much inferior to that of Kuznetsk. The beds cover about 90 sq. miles, and lie at a depth of more than 98 ft. The seams are horizontal with an average thickness of 9 ft. The railway is the chief consumer. Coal is used in the electric works in the town of Irkutsk, but not for heating purposes in private houses.

Yenisei and Lena Basins.—There are also considerable deposits in the Yenisei valley near Dudinka and in the Minusinsk region. There are said to be deposits in the Lower Tunguska and Chulim valleys. In the Lena valley coal has been located at a number of places, throughout its middle course and as far north as Bulun. On the west side of the Lena the deposits extend beyond the mouth of the Markha, a tributary of the Vilyui ; on the east side along the Aldan, beyond the mouth of the Maya. The coal, which is of recent origin, is not worked.

Transbaikal.—In the Transbaikal a number of coal deposits, mainly lignite, have been located ; along the shores of Baikal with outcrops near the lake, and in the valleys of the streams round, including the Uda and the Chikoi ; at seven points in the valley of the Khilok and at two near Lake Gusinoe. A seam near Pereemnaya, on the south-eastern shore of Baikal, is worked for the lake steamers, but the output is small. Large lignite beds exist in the valley of the Ingoda, one being at Novaya-Kuka, only 4 miles from the railway. Coal also occurs at several points in the valleys of the Shilka, Onon, and Argun. The deposits on the Argun have not been explored, but as the region where they occur is treeless, they may be of importance. About 1,500,000 tons are produced annually from these deposits along the Transbaikal Railway, but the quality is inferior, giving only half the heat of the Cheremkhovskoe coal.

Amur Province.—Coking coal of fair quality has been found about 30 miles from the mouth of the Dep, a tributary of the Zeya, on its right bank, but the amount seems to be small. In the Bureya valley several deposits of better quality have been found.

Extensive deposits of brown coal have also been located at several places along the Amur and the Amur Railway, notably near Khabarovsk.

Maritime Province and Sakhalin.—There is much coal, mostly of a poor quality, in the Maritime Province. Of 27 deposits 10 are being worked. The principal sources are a large lignite mine near Vladivostok, yielding upwards of 200,000 tons annually for the railway, a couple of small mines on Amur Bay, the Government mines at Suchan, 60 miles from America Bay, and the Dué, Alexandrovsk, and other mines in Sakhalin.

The Suchan mine, which supplies the Government requirements, is run at great expense and on non-economical lines. It is connected with the Ussuri Railway by a branch 93 miles long, but this can handle only 133,000 tons at most in a year. The output rose from 105,496 tons in 1908 to 206,783 in 1912. The quality of the coal for steam-raising has been much criticized and its calorific value is comparatively low. The coal is of three sorts, bituminous, anthracite, and coking, the percentage of volatile matter being respectively 27 to 30, 6 to 8, and 22 per cent. It was expected that the briquette works on Golden Horn Bay would turn out 3,200,000 briquettes in 1914. There are said to be indications of both brown and anthracite coal along the coast of the Maritime Province as far north as de Castries Bay.

A Government engineer, who has recently visited the Mongugai coal-field, which lies some 12 miles inland from the western shore of Amur Bay on the Mongugai River and is only 24 miles by land and water from Vladivostok, estimates that it contains some 5,000,000 tons of good anthracite coal, very similar to Welsh coal. The field is much cut up by eruptive rocks and heavily faulted. If a small harbour, with a narrow-gauge line, were built, 150,000 to 200,000 tons could be delivered annually at Vladivostok at a cost of about 10*s*. a ton, instead of the 18*s*. or more charged for Japanese coal, of which some 12,000 tons are imported into Vladivostok every year.

Mines near Alexandrovsk in Sakhalin are connected to the sea by a railway 7½ miles long.

The absence of a harbour has prevented the excellent Dué coal from being worked on a large scale. In 1912, 24,322 tons were produced from five pits. The Government is putting the mines up for auction for 36 years on condition that the lessee shall build a harbour for general use. The mines could certainly produce large quantities of coal at a moderate price if properly worked, as the seams are thick and lie horizontally, while timber is abundant. In quality it is said to be equal to Welsh coal. Excellent bituminous coal has also been discovered at the mouth of the Pilevo near the Japanese frontier.

Kamchatka.—There is a large deposit of brown lignite at Baron Korfa Gulf, north of Kamchatka. Similar coal occurs on the shores of Gizhiga and Penzhina Bays, as well as at several places on the west coast of Kamchatka and elsewhere in the Kamchatka province.

Salt

The Urals produce about 20 per cent. of Russia's salt, in spite of the fact that the industry there is not developing rapidly. The Orenburg deposits are among the richest in Russia, which contains some of the largest salt-beds in the world.

Salt is an important, but little developed, Siberian mineral. In the west it is chiefly found in a number of lakes in the Semipalatinsk territory, on both banks of the Irtish, in the Akmolinsk region and in the Government of Tomsk. The output varies considerably, as the deposits in the lakes depend on the weather. In 1911 the output of salt in western Siberia was 129,000 tons, one-sixth of the total output of the Russian Empire; in eastern Siberia about 10,600 tons.

In the Kirghiz steppe seven lakes are controlled by the Government and leased for working. All the other lakes of the region, many hundreds in number, are left for the use of the Kirghiz, but the salt they yield is less pure in quality. Most of them could only produce pure salt by the artificial basin system. Chief among the reserved lakes are the five west of Pavlodar, of which by far the most important is Lake Koryakov, 12 miles from the town. Its salt is considered the best in Siberia. In 1905 it produced over 32,000 tons. The method of procuring the salt is quite primitive, no machinery being used. Lake Karabas is another important salt lake, west of Semipalatinsk;

it supplies the needs of that town, as well as of Biisk, Zmyeinogorsk and Zaisan. This lake also belongs to the State. The salt is collected in the most primitive way and loaded on to camels.

In the Tomsk Government, in the Baraba steppes, are a number of salt lakes, of which Lake Burlinskoe, some 70 miles to the north-east of Pavlodar, is the most important. It produces by far the greater part of the output in this region. The salt is comparatively pure: it is used by the peasantry and in the fishing industry on the Ob.

Eastern Siberia abounds in salt, but the richest deposits of rock-salt and the best salt springs lie in districts too remote for profitable working. Most of the salt is obtained by evaporation from salt lakes, as, for instance, near Abakanskoe in the Yeniseisk Province. In this province evaporation produces 7,500 tons a year, about the same quantity as in the Irkutsk Province. The industry flourishes in Transbaikal, especially round Novi-Selenginsk, Ust-Kiranskaya, and Troitskosavsk. There are a number of brackish lakes, of which Baruntorei is the largest, in the southern part of the district. Lake Borzinsk near the Chinese frontier is the only salt lake in the Nerchinsk district, and the salt does not settle there every year. The most important saltworks in Siberia are at Ussolye near Irkutsk on the Lower Angara. They produced 10,000 tons in 1907. Work is only carried on in summer, owing to the cost of fuel. If the price of salt fell below about ½*d.* per lb., it could not be produced at a profit.

The Ust-Kutskoe works on the upper Lena have an output of some 1,600 tons annually for local use, but they are unfavourably situated. The salt is plentiful. Very large deposits of rock salt exist on the Vilyui and its right tributaries near Suntarskaya. The salt is contained in red clay and is everywhere accompanied by gypsum. On the right bank of the Kyundyaya, a right-hand tributary of the Vilyui, salt forms two masses in a mountain of red clay and gypsum. During the spring floods some of its tributaries, notably the Kampentszyaika, become quite brackish. At present there is no market for this excellent salt owing to the absence of means of com munication.

As many districts are totally devoid of salt, the Government keeps salt-depôts for supplying local needs in the remoter regions of Siberia.

Glauber's Salt.—In the northern portion of the salt lake system of western Siberia, i. e. on the Baraba steppe, the lakes always contain a considerable amount of salts other than common salt, principally sulphate of sodium (Glauber's salt).

In eastern Siberia Glauber's salt is produced from Lake Doroninsk in the Barguzin district, and also from the extensive deposits in the Minusinsk region, close to the left bank of the Yenisei. Lake Doroninsk likewise contains vast deposits of sulphate of soda, which is also produced from a couple of small lakes in the south of the Achinsk–Minusinsk region, the Kiransk Lake in the Transbaikal, and in the neighbourhood of Verkholensk on the upper Lena.

In the Tomsk Government hot mineral springs occur only in the mountainous districts. The Rakhmanovski and Byelokurikha springs, both in the Biisk region, near the Mongolian border, are the best known. The Transbaikal is rich in mineral springs, most of them cold. Very few of them have been exploited or even explored. They are especially numerous near the Shilka and in the Chita region. In Kamchatka mineral springs are likewise common, their healing qualities being often appreciated by the natives. There are sulphur springs on the middle Lena.

Precious Stones and Building Materials

Marble and lapis lazuli are common in the Irkutsk and Transbaikal regions. Very fine lapis lazuli has been produced in the valley of the Malaya-Bistraya, a tributary of the Irkut. Marbles, garnet, asphalt, &c., are found along the Talaya and Slyudyanka, which flow into Lake Baikal.

Ashirite, a rare kind of emerald, is found to the NW. of Karkaralinsk in the Kirghiz steppe and alabaster in the immediate neighbourhood of Semipalatinsk.

The Transbaikal is the richest district for precious stones. On the 'granite mountain' Adun-Chelonsk, near the confluence of the Onon and the Borzya, topaz, beryl, aquamarine, tourmaline, crystals, and other stones have been found. Very fine topazes come from the Nerchinsk Range, between the Unda, a tributary of the Skilka, and the Urulyungui, a tributary of the Argun. Neither of these districts is at present regularly exploited. Garnets in small crystals are also found on the Onon, 56 miles from Nerchinsk. On the eastern shores of Baikal, near Barguzin, there are fine rubies and almondines.

The basins of the Byelaya, a tributary of the Lower Angara, and its tributaries, the Iret and the Onon, contain nephrite, which is highly prized by the Chinese. Some of the blocks weigh 9 or 10 cwt. There is nephrite near Cheremkhovskoe.

Jasper exists near Zhigansk on the Lena. The jasper and porphyry of various colours from the Altai are celebrated. They come especially from the valleys of the Alei and Charish and are polished at the Kolivan works. From the eight quarries working in the Altai come porphyry, blue and green jasper, malachite, granite, marbles, breccia, smoky topaz, coloured quartz, agate, and chalcedony.

Near Olgi Bay is a mountain said to consist entirely of marble, to which a railway has been projected. Marble is also found on the upper Yenisei, on the southern and eastern slopes of Baikal, and in the basins of the Onon and the Argun.

Lime, building-stone, and common clays are found almost everywhere. Fire-clay and fire-resisting sandstone are worked near the mines in the Kirghiz steppe and the Irkutsk Government.

Kaolin and white clay for porcelain are worked in several places in the Irkutsk Government. Felspar and quartz for glass factories are obtained from deposits in the Baikal Mountains.

CHAPTER XIII

MANUFACTURING INDUSTRIES

Kustarni Industries—Factories—Chinese Industries in the Far East

KUSTARNI INDUSTRIES

THE characteristic industrial feature of western Siberia, as of European Russia, is the *Kustarni* (peasant or cottage) industry, performed in the houses of the cottagers, sometimes as the winter or nocturnal occupation of an agricultural folk during the hours of their unemployment, sometimes as the main occupation of the inhabitants. Several of the old settlers have given up agriculture for peasant industries, but many of these industries are decaying, and the railway by its distribution of the commodities accelerates their fall. Some patterns remain traditional in one family for generations, and the older these traditions, the better and finer is the work produced. Those who are responsible for developing these industries, try as far as possible to revive the finer and more artistic work which has had a tendency to give way before cheaper and newer products. In eastern Siberia these Kustarni industries are less developed, and the more easterly the province the less is the degree of development. In the Kirghiz steppes, where there is a considerable amount of peasant weaving, the Kustarni trades flourish more among the Kirghiz than among the Russian inhabitants. The only exception is a kind of tanning industry established by recent immigrants.

In western Siberia it was estimated about ten years ago that there were altogether 13,000 factories, employing 33,000 workpeople and producing about £4,000,000 worth of manufactures. As these numbers include the bigger factories in some of the towns it shows that only about two people in one house are employed on the same thing. Usually all the inhabitants of one village are employed in the same industry, and even all the villages on one road will have one or two staple occupations. So on the road from Kurgan to Yalutorovsk all the many villages that the traveller passes manufacture wool or leather. Some examples of Kustarni industries in western Siberia may be given. About Tyumen, a great centre of industry, the Kustarni manufactures are cooperwork, sieves, turned utensils, carts, furniture of a rough kind, sledges, pitch, the wooden parts of horses' collars, Russian ploughs, and in the Kamen volost of the same district carpet making, with bright flowers and animal patterns. The wool, colours, and designs are bought from traders—formerly vegetable colours were got from the Samoyedes, now aniline dyes are used: the Russians love bright colours. Round Turinsk and Tobolsk much carpentering work is done, but in Tobolsk fishing nets and carvings from mammoth ivory are also made, and in Turinsk anchors because of the fishing. Throughout south Tobolsk wool and skin products are manufactured. At Kurgan and Turinsk they make plough-shares, at Ishim ropes and rough agricultural machinery, at Samarovskoe leads for the fishing-nets. Throughout the Tobolsk Government they make plaited bast-work, harness, skin boots, wool products, axles, distaffs, and troughs. In the Tomsk Government at Kuznetsk and Tomsk there are smiths and joiners, at Barnaul they make metal pots for milk and the special skin coats known as 'barnaulkas', and at Biisk cedar-nut oil.

The Ministry of Agriculture has tried to improve the technical knowledge of the peasant-workers by establishing educational workshops, of which there are now seven in the Tomsk Government, giving instruction in such things as weaving, furniture-making, carpentering, the manufacture of agricultural machines, cart-building, pottery, and tanning.

In eastern Siberia labour is scarcer and dearer, and the standard of workmanship is much lower. In Irkutsk and

Yeniseisk woodwork of various kinds comes first in importance, especially in the neighbourhood of the big towns ; second comes the dressing of sheepskins and wool products ; then weaving and metalwork, which are much less advanced About Yeniseisk there is pottery, and near Irkutsk boot-making. In Transbaikal coopers' work alone is of importance. In Yakutsk one special form of Kustarni industry should be mentioned, the fine work done in mammoth ivory by the Yakuts. In the Amur and Primorsk Provinces cottage industries hardly exist at all, but the government is doing its best to organize them, as offering excellent employment, especially in the remoter districts, during the winter. Instructors are brought from Russia, schools are opened, and assistance is given in procuring raw material and organizing sales. It is thought that wooden articles for local use, such as furniture, sledge runners, &c., could be made at home. Charcoal burning, the dry distillery of wood products and the making of rough tools and pottery might be taken up, as well as weaving, coopering, tanning, and shoemaking. In the Primorsk seven workshops and schools have been or shortly will be established. In Amur and Primorsk £23,000 was to be spent in 1914 to promote these industries.

Factories

In the towns bigger factories have grown up, some of which have been established for 70 or 80 years, as, for instance, the celebrated glass factory 28 miles west of Krasnoyarsk, established about 1840 and employing normally 400 and at special seasons 800 workmen. It is difficult to define very clearly the distinction between the Kustarni and bigger factories ; very often the same industry is carried on in both, or a factory may be merely an aggregation of Kustarni workers. The chief occupations of Siberian towns are distilling, brewing, tanning, soap and tallow factories, flour-milling, saw-milling, weaving, oil-milling, rope-making, glass-making, brick-making, pottery. In the Russian Far East the managers of factories are usually foreigners, especially Japanese, until the Russo-

Japanese war. Skilled labour was supplied by the Chinese, unskilled by Koreans. The retail trade was chiefly in the hands of the Chinese.

Distilling and Brewing

The distillation of spirits, especially vodka, is the oldest and most firmly established industry in Siberia. Vodka distilleries supply only the state, which has its own retail shops, and export a limited amount to other countries. There are about 60 distilleries in Siberia, of which the most important is at Semipalatinsk, in which in 1906–7 6,000,000 litres were produced.

Other important distilleries are in Kurgan, Tomsk, and the neighbourhood of Omsk. Spirits are distilled from rye, wheat, and potatoes. Before the prohibition of vodka Siberia could not satisfy its own needs under this head, but imported from Russia. There is a Kustarni industry in the distilling of *samosidka* (a kind of vodka) from corn, but this is only for local consumption.

Breweries before the war were usually in the hands of Germans and Austrians. The chief kind of beer manufactured was a light lager beer of German type. Nearly all the towns of any importance have their breweries, producing between 600,000 and 1,200,000 litres per annum. The work is usually done in winter. The beer produced is of fair quality, like the ordinary Russian beers, but with a somewhat higher proportion of alcohol. Besides beer, mead and kvas are brewed. There are yeast factories at Omsk and Irkutsk.

Animal Products

Tanneries work both for local use and export. The largest, which are at Tyumen, work some 60,000 cattle hides and 10,000 horse hides per annum. The next largest are at Biisk. The chief supply of raw hides for western Siberia comes from Kurgan, Semipalatinsk, Barnaul, and the great market for these commodities at Petropavlovsk. There are no tanneries in Mongolia ; so a considerable amount of finished products go into that country, some of which had previously been sent out of it as untanned hides. This constitutes much of the trade which passes from Biisk along the Chuya track. A resident at Kyakhta has bought a Tyumen tannery in order to facilitate the trade through Kyakhta. Other places where tanning is important are Petropavlovsk, Tomsk, Novo-Nikolaevsk, Barnaul, Zaisan, Omsk, Semipalatinsk, Zmyeinogorsk, and Irkutsk. The Lena goldfield is largely supplied by leather from Ussolye near Irkutsk. Tanning is a frequent Kustarni industry as well.

Tallow, soap and candles are a common industry, especially in Tobolsk and northern Akmolinsk. Export is made by rail to Russia, especially to Moscow, Petrograd, Kazan, and Urakovo. Ninety-three per cent. is exported in winter, for the packing expenses are less heavy then, as the use of casks is not obligatory. Petropavlovsk and Kurgan are the chief places of export.

The Kirghiz cattle supply the material. The soap and candle factories supply the commonest sorts for local needs. For superior wax candles the church has a monopoly with factories in Tobolsk and Krasnoyarsk. The principal soap works are in Petropavlovsk, Omsk, Tomsk, and Blagovyeshchensk.

Other industries connected with animal products are wool-dressing, especially at Tyumen and Irkutsk, and weaving. The carpets of Tyumen have been already mentioned.

Timber Products

The saw-mills of Siberia (a very important industry) are described in Chapter XI. The carpentering and coopering are mainly Kustarni work. A paper factory in a village near Tyumen employs about 300 workpeople and has an output of about £30,000 worth. There is boat-building on the Tavda, at Tyumen and at Minusinsk. Carriage-building is an industry at Tomsk, and to some extent at Omsk and Irkutsk. There is an establishment for the preparation of railway sleepers near the railway station of Omsk.

There are several match factories at Tobolsk, Tomsk, Tyumen, and in the Biisk district. The Tomsk establishment is the best; there 300 hands produced daily 350 chests or 32,000 boxes (1903). The other places employ about 100 hands. The phosphorus (where used) is obtained from works in Perm, the other materials being of local manufacture. Siberia provides for itself in the match industry, successfully competing with Japan in the far eastern regions.

Metal Foundries and Engineering Works

Along the railway line are engineering works and works for repairing the railway. Next to the railway works the best equipped workshops are those attached to the technical high school in Tomsk and some industrial schools in Omsk and Irkutsk. River steamers and engines are built at Tyumen and Blagovyeshchensk. There is a bell-foundry at Tyumen, a brass-foundry at Barnaul, a nail factory at Irkutsk, and a factory of agricultural machinery (the superior kinds are imported) at Omsk. In eastern Siberia there are two iron and five copper foundries at Blagovyeshchensk, besides mechanical factories at Khabarovsk and Nikolsk-Ussuriski.

Glass and China

Bottle-glass, window-glass, and rough table-ware are the only glass manufactured in Siberia, the superior kinds being imported. The chief glass factory is the one mentioned above west of Krasnoyarsk, the workmen of which are Europeans, not Siberians. There are others near Kurgan, Biisk, Tomsk, Minusinsk, and Irkutsk. Some of these employ about 100 workpeople. In the east there are two glass factories, one in the Amur Province and one in Nikolsk-Ussuriski.

There is one large china and porcelain factory in Siberia, at Khaita, on the Byelaya, about 90 miles from Irkutsk. It produces some 483 tons of less fine ware for Siberian use only. A Moscow firm supplies most of Siberia.

Brick, Pottery, and Cement

Brickworks are very general near the Siberian towns. The brick factories are often worked by steam. Although the Siberian houses are built principally of wood, the foundations are generally of brick, and so there is a larger demand for bricks than might be expected. Prices range high : often about £1 for 100. In 1907 the principal steam brick-works were two in Tomsk, which produced three million bricks a year each, two at Krasnoyarsk, working much below their capacity, two at Irkutsk, which produced one to two million bricks, but of which one had suspended work. Other important brick-works were in Omsk, Novo-Nikolaevsk, Cheremkhovskoe, and Aleksandrovskoe ; at these bricks were made by hand.

Pottery is little developed. The peasants replace it with numerous substitutes of wood and bark, while blue and white enamel ware (Austrian and Polish) has considerable vogue. There are, however, potteries near Tyumen. The most important pottery is at Polovinnaya on the Byelaya, 64 miles west of Irkutsk, where boats can reach the factory. The clay here is good. Cement works are found in Nizhne-Udinsk, where one produces about 50 tons annually. Two others in Transbaikal manufacture for local needs, and another was established in 1907 near the Ussuri Railway at the cost of £100,000.

Oil

Another large industry is the manufacture of vegetable oil, the milling of which goes frequently with that of grain. The oil is mainly from flax and hemp. Among the principal oil mills are one at Ust-Kamenogorsk for sunflower seeds, and one at Petropavlovsk, for linseed (500 tons per annum). Mills abound all over western Siberia ; from the Urals to the Ob they are mostly windmills, east of the Ob they are mostly watermills. In 1910 there were 1,041 oil-mills in the Tomsk Government.

Miscellaneous

A few important factories not otherwise classified may be mentioned. In Irkutsk and Khabarovsk are cigarette-case factories; Irkutsk contains ten printing works (there are others at Omsk and Tomsk), two steam sausage manufactories, and a pearl-barley factory. Barnaul has a soda factory and a considerable rope factory with an output worth £2,500 a year, and there is another at Minusinsk. Usually rope-walks are a Kustarni industry. At Kolivan there are stone-polishing works which belonged to the Cabinet of the Tsar. Vases are fashioned of marble, jasper, and breccia, but the conditions of the industry are said to have remained primitive.

Near Posiet Bay, Olgi Bay, and other bays in the Maritime Province seaweed is the basis of a valuable iodine industry. An iodine factory has just been built at Nakhodka Bay near Vladivostok, and in 1916 it was expected that 8,000 tons of the weed would be dealt with. It is now to be cut with hooks and not torn up by the roots.

Chinese Industries in the Far East

An edible seaweed, variously named, sometimes called 'sea colewort', is collected in the south of Primorsk, whence some 1,600 tons are annually sent to Japan and China, and in Sakhalin, whence some 2,700–3,200 tons are annually exported. The quality depends on the drying, but as this must be done on stony ground, and as three days of sun without moisture are essential, the process is not easy. There are three qualities: the best from the Maritime Province goes to Shanghai, the rest to Chefoo. Mention has been made of the *panty* made from deer horns (see p. 56). To this should be added the gentian root which is found in the depths of the virgin forests along the tributaries of the Ussuri, and which is supposed by the Chinese to possess marvellous properties as a medicine. It sells according to its age, the older the more valuable, at £5 to £23 a pound. The Chinese have attempted to grow it in Russian territory, but the cultivated root sells for only 25*s*. a pound. America meantime exports £160,000 worth of the root to China annually.

CHAPTER XIV

NATURAL RESOURCES AND TRADE OF ARCTIC RUSSIA

Timber—Agriculture and Stock-breeding—Mineral Wealth—Fur and Eider-down—Industry and Trade

Timber

The most valuable asset of northern Russia in Europe is its forests. In the Government of Arkhangel, in 1897, it was estimated that out of 211,356,000 acres 81,000,000 were forest—the waste land amounted to 129,000,000 acres, while only 756,000 acres were available for agriculture. Along the Murman Railway there are great timber areas, mostly spruce and pine. As there is a good supply of water-power and 'white' coal in the neighbourhood, these could be profitably exploited. Sawmills are the principal industry of this government, but their number might be indefinitely increased, and the material sawn up might be exported in its manufactured form: at present it is mostly logs and sawn wood that are exported. The carpentering that is done in the country only meets the needs of the local population (see p. 132). Sawmills have been erected at Keret and at other settlements along the White Sea coast, and also on Kola Inlet. Large timber concessions on the Pechora River were made in 1916.

A cognate industry is the manufacture of pitch and tar, the most important occupation in parts of the Shenkursk district and in other parts of this government. This might be usefully developed. At present local conditions differ considerably; the rate of taxation on the possessor of the woods varies, but is usually too high, especially in the extreme north, where the period of production is shortest. In the state lands of Arkhangel the peasants have certain privileges,

e. g. free use of wood for heating the 'kettles'; in some parts the by-products are taxed. When the country is deforested by the tar industry, care should be taken that the cleared land is used for agriculture, and not allowed to decline into marsh-land. At present a large part of the population are employed on this industry, but it needs fostering by supplying means of transport, lowering the taxes, arranging for a system of credit, introducing technical improvements and experimental stations, and securing that the wood left over is utilized in the sawmills of the north coast region. There are extensive works in the district of Velsk between Arkhangel and Vologda, where special attention is paid to the production of tar. Pitch is exported principally to the United Kingdom from Arkhangel.

Agriculture and Stock-breeding

Agriculture is only remunerative in the Shenkursk district near the Vaga, in the southern portions of Kholmogor, Onega, and Pinega, and on the 'summer' shore of the south coast of the White Sea. The principal crop is barley, to which a large proportion of the farm lands is devoted, and which is probably three times as extensive as rye, the next largest crop, which grows, however, up to Askino in the estuary of the Pechora. Oats and buckwheat are grown less extensively. The local harvests, however, only meet the needs of one-third of the population. Other crops of this region are potatoes, peas, beans, cabbage, turnips, carrots, spinach, sunflower, hemp (both mezenskaya and a larger kind), and flax. The Juravski expedition of 1909 reported most favourably on the prospects and present condition of agriculture in the northern Pechora district, adding that bad harvests there had never been remembered by the peasants. There is a floating grain-elevator at Arkhangel, the only one in northern Russia.

Cattle-breeding could be made very successful. The Kholmogor breed, established by Peter the Great, by interbreeding the local cattle with others introduced from Holland, is famous for its size, beauty, and milk-producing powers. The alluvial meadowland along the Northern Dvina, Mezen, and Pechora is good for cattle, and at small expense the swamps of the Arkhangel Government could be turned into land covered with rich grass. Of the 756,000 acres available for agriculture 540,000 were pasture land in 1897. There were at that date in the Government of Arkhangel 114,962 cattle, 145,590 sheep, 52,109 horses. It is estimated that the tundras of this region feed nearly 300,000 reindeer.

Mineral Wealth

The mineral resources of the country are little developed. As long ago as the chronicles of 1558 gold was recorded in the sand of the Rivers Kola and Tuloma. There are traces of silver, lead, and zinc everywhere on the Murman coast, and iron and copper are present elsewhere in the Kola Peninsula. Copper used to be worked centuries ago on the Tsilma, a tributary of the Pechora. Ores with 80 per cent. of lead have been found on the River Varsina. Coal is found near Onega ; naphtha in the Pechora district, and the Utkinski oilfields in the Cherdin district of Arkhangel are said to resemble those of Pennsylvania; a railway projected from Bogoslovski to the White Sea will open them up. On the 'summer' coast of the White Sea there is found a rich clay for porcelain of a bluish-white colour.

Fur and Eider-down

The fur trade has decreased considerably, and in 1902 serious measures had to be taken in order to preserve what was left of certain valuable animals, e. g. the ermine and the Arctic fox. In 1897 there were killed in the Arkhangel Government 386,771 wild animals, and 242,666 brace of birds, with a total value of £12,388. Much the largest number of animals were squirrels, viz. 384,189. In comparison the rest were negligible. Second in order came foxes, of which only 1,092 were killed.

Eider-down is obtained from Ainov Island, where the nesting-places of the ducks are protected.

Industry and Trade

Most of the industrial occupations of the inhabitants have been already recorded. Besides these should be mentioned the tallow industry which is strongest in the Kola district, and the chamois leather industry, which is strongest in the Pechora district. In 1897 the manufactures of the Arkhangel Government were worth £773,160. Among the industries were shipbuilding, leather, canvas, cordage (though most of the hemp is exported for the fleets of other countries), nets, and potash.

CHAPTER XV

DIVISIONS, ADMINISTRATION, AND TOWNS

Siberian Boundaries—Governments and Territories—Administrative System—Administrative Districts—Siberian Towns—Local Government

Siberian Boundaries: Political and Geographical

The governments into which the Russian Empire is divided disregard the geographical boundaries of Siberia. The governments of Perm and Orenburg, which are European, cross the Ural Mountains and include about 150,000 square miles of Asiatic Russia, and the south-west portion of Siberia is included in the general government of the Steppes (governments of Akmolinsk and Semipalatinsk), which is administered from Omsk, and the northern portion of Turgai, which belongs to Central Asia politically, would geographically be assigned to Siberia.

Governments and Territories

The grouping of the administrative districts has constantly changed and further changes are imminent. The old division into Western and Eastern Siberia is given up; Western Siberia used to consist of the governments of Tobolsk and Tomsk, while the other administrative divisions constituted Eastern Siberia. Now in all Siberia there are four governments (*Guberniya*), Tobolsk, Tomsk, Yeniseisk, Irkutsk; eight territories or provinces (*Oblast*), Yakutsk, Transbaikal, Amur, Maritime (Primorsk), Kamchatka, Turgai, Akmolinsk, Semipalatinsk (of which the last three belong to Central Asia and the Steppes), and one division, Sakhalin. Some of these are included under general governments: the general government of Irkutsk includes Irkutsk, Yeniseisk, Transbaikal, and Yakutsk; the general government of the Amur (Priamur) includes Amur, Kamchatka, Maritime, and Sakhalin, and the general government of the Steppes includes Akmolinsk and Semipalatinsk.

The old Maritime Province is now divided at about 56° N., the northern part stretching thence to the Bering Strait being called Kamchatka; it has also been decided to transfer the residence of the governor of Sakhalin (i. e. that part of the island north of lat. 50° N. which Russia has retained since the Treaty of Portsmouth, 1905), to Nikolaevsk, and to form a new district to be known as the Nevelski district, including together with northern Sakhalin the central part of the Udsk district in which Nikolaevsk lies. Also the boundaries of the Amur province are to be enlarged by the addition of the south-east portion of Transbaikal, and two new districts (Selemdzha and Zeya) are to be formed.

Administrative System[1]

Tobolsk and Tomsk are administered on the same system as the Governments of European Russia, and the Eastern provinces will be administered similarly, with civil governors taking the place of military governors except in Kamchatka. Where the European system prevails there is a general administrative council, presided over by the deputy-governor, directly under the authority of the Minister of the Interior. Over it presides a civil governor representing him, who is assisted by councillors nominated locally and approved in Petrograd. Usually the business is deputed to committees as below:

(1) Under the Minister of the Interior—Department for (*a*) Urban Affairs, (*b*) Peasant Affairs, (*c*) Justice; Prison Committee, Education Board, Land Valuation Staff, Public Health Department.

(2) Under the Imperial Finance Minister—A local branch of the Imperial Treasury for assessing local and imperial taxes.

(3) Under the Minister of Agriculture—Agricultural Organ-

[1] At the time of the Russian revolution in 1917.

ization Committee, Local Immigration Committee, Land Survey Staff.

(4) Under their respective central bureaus—Ministry of Trade and Commerce, Local Department of Justice, Local Military Authority.

There are eight members elected to the Imperial Duma from Siberia by indirect vote.

Each province is divided into *uyezds* (districts), each *uyezd* into *volosts*; each *volost* into villages, those with a church being styled *selo*, those without a church being styled *derevnya*. Over the *uyezd* presides a *uyesdi nachalnik*, over the *volost* a *zasidatil*, over the village a *starosta*.

Administrative Districts

Tobolsk. Pop. (estimated 1913), 2,005,000. Area, 539,659 square miles. Capital, Tobolsk. Districts: Tobolsk Tyumen, Kurgan, Tara, Ishim, Tyukalinsk, Berezov, Surgut, Turinsk, Yalutorovsk. Police divisions, 30; volosts, &c., 270, rural communities, 2,609; other settlements, 4,760.

Tomsk. Pop. (estimated 1913), 3,919,000. Area, 331,159 square miles. Capital, Tomsk. Districts: Tomsk, Barnaul, Biisk, Kainsk, Kuznetsk, Mariinsk, Zmyeinogorsk. Police divisions, 31; volosts, &c., 294; rural communities, 3,194; other settlements, 3,350.

Barnaul is the head-quarters of the Altai Administration and the centre of the 'Cabinet' estates (i. e. belonging to the Tsar).

Irkutsk. General Government. Capital, Irkutsk.

Irkutsk. Pop. (estimated 1913), 733,000. Area, 287,061 square miles. Capital, Irkutsk. Districts: Irkutsk, Balagansk, Kirensk, Nizhne-Udinsk, Verkholensk, Vitimsk. Police divisions, 25; volosts, &c., 116; rural communities, 579: other settlements, 2,336.

Yeniseisk. Pop. (estimated 1913), 982,000. Area, 987,186 square miles. Capital, Krasnoyarsk. Districts: Krasnoyarsk, Achinsk, Kansk, Minusinsk, Yeniseisk, Turukhansk, Usinsk. Police divisions, 21; volosts, &c., 103; rural communities, 1,639; other settlements, 1,464.

Yakutsk. Pop. (estimated 1913), 328,000. Area, 1,533,397 square miles. Capital, Yakutsk. Districts : Yakutsk, Olekminsk, Sredne-Kolimsk, Verkhoyansk, Vilyuisk. Police divisions, 15 ; volosts, &c., 47 ; rural communities, 383 ; other settlements, 337.

Transbaikal. Pop. (estimated 1913), 920,000. Area, 236,868 square miles. Capital, Chita. Districts : Chita, Barguzin, Aksha, Nerchinsk, Nerchinski Zavod, Selenginsk, Troitskosavsk, Verkhne-Udinsk. Police divisions, 25 ; volosts, &c., 141 ; rural communities, 951 ; other settlements, 791.

Amur (Priamur). General Government. Capital, Khabarovsk.

Amur. Pop. (estimated 1913), 241,000. Area, 172,848 square miles. Capital, Blagovyeshchensk. The government is military and divided into Cossack regiments and battalions. Police divisions, 7; volosts, &c., 23 ; rural communities, 325; other settlements, 259.

Maritime Province (Primorsk). Pop. (estimated 1913), 604,000. Area, 295,664 square miles. Capital, Vladivostok. Districts : Khabarovsk, Udsk, Ussuri Cossack, South Ussuri (Nikolsk-Ussuriski), Iman, Olgi. Police divisions, 19 ; volosts, &c., 47 ; rural committees, 941 ; other settlements 1,031.

Kamchatka. Pop. (estimated 1913), 39,000. Area, 503,777 square miles. Capital, Petropavlovsk. Districts : Petropavlovsk, Okhotsk, Gizhiga, Commander Is., Chukotsk (Chukchee), and Anadir. Police divisions, 4 ; settlements, 254.

Sakhalin. Pop. (estimated 1913), 14,000. Area, 15,334 square miles. Capital, Alexandrovsk. The province is divided into two districts, Alexandrovski and Timovsk. Police divisions, 4 ; rural communities, 35 ; other settlements, 35.

Steppe. General Government. Capital, Omsk.

Akmolinsk. Pop. (estimated 1913), 1,492,000. Area, 229,609 square miles. Capital, Omsk. Districts: Akmolinsk, Atbasar, Kokchetav, Omsk, Petropavlovsk. Police divisions,

26; volosts, &c., 223; rural communities, 1,182; other settlements, 579.

Semipalatinsk. Pop. (estimated 1913), 862,000. Area, 184,631 square miles. Capital, Semipalatinsk. Districts: Semipalatinsk, Karkaralinsk, Pavlodar, Ust-Kamenogorsk, Zaisan. Police divisions, 20; volosts, 163; rural communities, 1,094; other settlements, 608.

Of the other provinces of the Steppe General Government, Uralsk does not concern Siberia, and Turgai does only in its north-east portion.

The district of Uryankhai, round the upper waters of the Yenisei, is nominally part of Mongolia, but is 'under the Russian sphere of influence'.

Siberian Towns

Origin.—The old towns in Siberia are either, like Yalutorovsk, built on the site of some Tartar city, or, like Irkutsk and many others, they were originally stockaded forts erected to keep the natives quiet and to form the centre to which they brought their 'yassak'. Other towns have owed their origin to special conditions: so Yekaterinburg grew up in the eighteenth century as the centre of the Ural mining district, and more recently Bodaibo as a mining centre in the Lena country and Novo-Nikolaevsk as the place where the Siberian railway crosses the Ob.

Appearance.—Siberian towns are seldom of an impressive appearance. Most of the private houses are of one story and of wood, though a disastrous fire in Irkutsk in 1879 led to an order that all further buildings in that city should be of stone, so that it has a more distinctive appearance. The public buildings and churches are usually of stone, and are often well built, and some fine cathedrals have recently been built. A common feature in Siberian towns is the triumphal arch, usually erected to commemorate the visit of some Tsar or Tsarevich. The bigger towns possess a number of schools, and there are several good libraries and museums. The bigger towns all possess theatres, and Krasnoyarsk has a very fine park. There are often good private residences of brick or stone. The shops, even when well stocked, make a poor display, as it is not customary to show the goods in the window. With but few exceptions the hotels are poor, except for the cuisine, and travellers have to bring their own bedding.

In the older towns the centre often has irregular and winding streets, but the rest of the town is laid out with extreme regularity in parallelograms. The streets are broad : paving, where it exists, is of wood, and the sea of mud in the middle of the street engulfs (as at Khabarovsk) the cement with which it is attempted to give it a surface. The station is often at some distance from the town, and in its vicinity a considerable settlement usually springs up.

Industry and Commerce.—Siberia is not a manufacturing country ; many of the factories, even in the towns, employ merely three or four hands ; but factories on a bigger scale are developing in some of the western towns. The commonest industries are tanning, soap-boiling, brick-making, milling, brewing, and the distilling of brandy or vodka. Near the mining districts there are also smelting works and laboratories for the assay of precious metals. Some towns owe their importance to the exchange of goods, and there are places especially where Europe and Asia exchange their goods as in Kyakhta, where the teas and furs of China are brought to the great exchange courtyard, and Petropavlovsk, where the materials and Asiatic goods of the Kirghiz steppe are brought to the historic Barter Court. The more southerly towns have weekly bazaars, and others have annual, or, more frequently, biennial fairs of great importance.

All important towns are described in the chapters on the various rivers in Volumes II and III.

Local Government

There have been since 1894 municipal Dumas in certain of the towns. In these the citizens elect a town council (*gorodskaya duma*) whose term of office is four years ; and this in turn appoints a board of aldermen (*gorodskaya uprava*). The former is deliberative and legislative in its functions, the latter deliberative. Both boards are under the presidency of the mayor (*gorodskoe golova*). The electors are all who pay the apartment-tax (*kvarterni nalog*).

The Mir.—The institution of Zemstvos, or provincial assemblies, has never been extended beyond European Russia, except in the Asiatic part of Perm. The great difficulty has been that Siberia is almost exclusively a country of peasants, and contains no educated or landowning class. The proportion of inhabitants who cannot read is very large, and to such a community an institution like the Zemstvo has been regarded as inapplicable. But since the emancipation of the serfs, in 1861, the *Mir* (=' world'), or folk-moot of the village community (a very ancient institution), has been organized, and has had self-government. It is the assembly of the peasants, and no one except a peasant can have a vote in it. With this body rests the allocation of village lands. It assigns to each family a hut and yard and a suitable amount of land, taking care that there is a proper proportion of arable, pasture, and forest-land. The average amount for each family is about 40 acres, but an additional amount is assigned to each male member. A new division is made every fifteen years, and a majority of two-thirds can re-allot land. Since 1906 a peasant can hold his land in perpetuity. A certain amount is always set aside for common pasturage. The chairman (*selski starosta*) is elected annually—in some places triennially—by the male inhabitants, the widows, and the wives of absent male residents. The decisions of the *Mir* are allowed to be unanimous, a defeated minority withdrawing its opposition. The *Mir* has also judicial rights in petty cases.

Powers of the Mir over individuals.—It is not usually an enlightened body, and it is apt to be a check on progress. Its traditional wisdom is likely to set itself against unfamiliar experiments in farming, and it puts obstacles in the way of free movement about the country, for it is afraid of losing the proportion of the taxes paid by the individual who wishes to migrate. It used even to send for members of the community who had thriven in the towns and compel them to return under an armed escort to their native villages. Those who do migrate do not cease to be members of their *Mir*, and frequently one who has prospered in town life continues his membership of the village community and the payment of his quota of the taxes. Taxes are low; the total of the direct peasant-taxes paid in the whole of the Yeniseisk Government amounts to £90,000, which works out at three shillings per head per annum. Custom allows the *Mir* to prevent any one it chooses from sharing in its debates, and the law permits this, provided such ostracism is not enforced for more than three years. But there is a more extended power that it exercises, for it can banish any undesirable person from the community, which usually means in European Russia that such a one is sent to Siberia. In many years half the prisoners sent to Siberia have been so banished.

Volost courts.—The *Mirs* are combined into *volosts* (cantonal assemblies) which elect an elder (*starishna*) and small tribunals (*volostnye sud*) for settling certain civil and criminal cases. The *starishna* (like the *starosta* in the *Mir*) is assisted in his decisions by a *pizar* (secretary), who often becomes the leading authority in the village, as the one person in an unlettered community who has any education at all. The *pizar* is usually out of touch with the peasant class by birth, habits, and education; he is indifferent to their welfare, and, being ill-paid, is constantly on the look-out for means of improving his position. While the *muzhiks* (peasants) wear the national dress, the *pizar* emphasizes his importance by wearing 'German' dress. In 1911 an attempt was made to get rid of these volost courts; they were eventually retained, but put under the supervision of local boards of magistrates, who constitute in this instance courts of appeal. In the *volost* the vote is not necessarily unanimous, but there may be a majority. What is dispensed in these courts is patriarchal justice based on customary law. Civil cases involving less than the value of 100 roubles come within their competence, as also do more important cases, provided that both parties to the suit agree to such an arrangement. They can exact fines up to 3 roubles or inflict seven days' arrest. They are also charged with maintenance of order in the *Mir* and in the family. There is no appeal (other than in cases of banishment) against the volost-court, unless it has acted *ultra vires*. Such appeals are lodged before the assembly of rural surveyors.

Other forms of lower courts.—Besides the volost-courts, which apply mainly to the normal Russian settler, there are other varieties of lower courts. In the military lands of the Cossacks there are *stanichni* courts, which deal not only with the affairs of the Cossacks, but with those of the tribes under their jurisdiction. The courts operate under the jurisdiction of the atamans of their districts. Appeals are lodged before the Cossack 'Provincial Economic Administration of the Cossack Armies'. The judges are elected by the Cossack assembly. In the Steppes the *narodni* court proceeds according to customary right (*adat*) or the written Musulman law (*shariat*). For the latter there exist special Musulman lawyers (*mufti*). These courts exercise extensive jurisdiction in civil and criminal affairs: e.g. they can punish theft with imprisonment for a year and a half. The *narodni* judges (*kazi-bii*) are chosen at the three-yearly sessions of the volost delegates. They must be thirty-five years of age and must have certain educational qualifications. In the Buryat districts of the Transbaikal Province there are special tribal courts, which settle the affairs of the Buryats on the basis of existing Buryat customs, which are founded on the ground of old Mongol steppe laws. They have jurisdiction in certain kinds of civil cases, where there is no limit to the amount of claim; in others the limit is assigned at £200: in criminal cases they can inflict fines of £30 or imprisonment of six months for a first offence. There is an appeal against their decisions to the rural surveyors. They were established in their rights by a law of 1901. Among the more vagabond nomad tribes there are so-called verbal courts (*slovesnaya raprava*), in which the procedure is by word of mouth. Another system, called the Turkoman system, is employed in central Asia.

Bureaucratic control of local government.—These democratic elements are counterbalanced by restrictions from a central and bureaucratic authority. There is a power of general supervision lodged in the 'district committee for the affairs of peasants'. From 1899–1912 there were rural supervisors (*zemski nachalniki*), but their place has now been taken by the reformed cantonal courts, indirectly elected by the *Mirs*, whose jurisdiction is confined to peasant cases. The *zemski nachalnik* among other functions used to administer the *zemski smet*, i.e. that proportion of the imperial taxes which is allotted by the Imperial Treasury for such purposes as roads, bridges, &c. In those towns which have municipal institutions this is administered by the town council.

Principal officials.—The governor of the province appoints over each district (*uchastok*) an official (*ispravnik* or *uyesdi nachalnik*) who acts as his representative and is the local commissioner of police. He in turn appoints his commissaries (*stanovoi pristav*), who are a sort of subordinate police-officials, acting as an alternative to the volost-courts for those who care to bring their cases before them; but this often involves travelling some distance, and, even when they are accessible, they are usually ignorant and illiterate men, without any knowledge of the law. He also appoints the *krestyanski nachalnik*, who collects the taxes, and the *mirabui sud* or justice of the peace. There are appeal courts constituted of these latter, together with two representatives of the volost-court. There is a justice of the peace for every three or four volosts.

Information about the government of the various native tribes will be found in the chapter that deals with them, under the headings of the separate tribes.

CHAPTER XVI

ROADS AND TELEGRAPHS

Roads and Travelling—Sledging—Telegraphs and Cables—Wireless Telegraphy.

Roads and Travelling

The Trakt and other main roads.—The great military road of Siberia, called the *Trakt,* was definitely constructed early in the eighteenth century, though there are references to such a road as early as the sixteenth century. Four main roads converge on Omsk, passing through the Urals by way of the towns of Verkhoture, Yekaterinburg, Zlatoust, and Verkhne-Uralsk: from Omsk the great road passes by way of Tomsk and Krasnoyarsk to Irkutsk, Chita, and Stryetensk. The old important route of the tea trade goes south into China from Verkhne-Udinsk, while beyond Chita an inferior road continues to Blagovyeshchensk after which a better road goes to Khabarovsk. Only very bad roads go on to Nikolaevsk and Vladivostok. Other important roads branch off to the north from Tulunovskoe and Irkutsk to Yakutsk and beyond, to the south from Omsk to Semipalatinsk and the Chinese frontier via Kosh-Agach, from Gryaznukha to Barnaul, and from Achinsk to Minusinsk. Along the lines of the Rivers Lena and Amur the roads are a less popular method of communication than the rivers, which are navigated in summer by steamers and used by sledges as frozen roads in winter, the posthouses being often along the banks of the rivers and away from the roads.

Seasons for travelling.—The worst seasons for travelling in Siberia are the autumn, when the frosts are beginning, and the spring when they break up. These two seasons are called in Russian *rasputitsa.* In summer travellers in the steppe districts suffer much from the dust, which often rises above the axle-trees, and is ready to be converted into a sea of mud by the rain ; in winter the travelling is mainly by sledge, and the frozen rivers add available roads. The Lena and the Amur are staked with double rows of pine branches to indicate the tracks.

Posting.—Along all the main roads are posting-houses at irregular intervals, usually about 12 to 20 miles apart. There has long been an efficient posting system in Russian Asia ; in fact an effective organization of the posts was made in Siberia earlier than in Europe. The Government appoints the post-masters, and they are allowed to make such terms as they choose, but the ordinary tariff is 1½ kopeks the verst for each horse in western Siberia, and 3 kopeks in eastern Siberia and the north generally ; in addition there is a Government tax (*pogron*) of 10 kopeks per horse on each stage. Among the nomad tribes the traveller has to use tents (*yurtas*) instead of posthouses, and the route will change according to the season of the year. There is no fixed rate of payment, but the charges depend upon the local Russian official. The posthouse is usually the nucleus of a small population, the position of which it has determined. Those who breed and provide posthorses are exempt from the imperial poll-tax (17 roubles). At each posting station 15 to 30 horses are kept, and about one-third of that number of tarantasses, *provoloki* (two-wheeled cars), sledges and drivers in attendance. It is customary for travellers to purchase their own vehicles for the journey: sometimes if they are fortunate they can sell them advantageously at the end of their route. Thus in places beyond Lake Baikal, like Stryetensk, where iron axles are not manufactured, it is easy to sell at a profit a tarantass which possesses these advantages. But a traveller from the east to the west is not likely to make a good bargain, when he sells his carriage at the end of his journey. Those who do not travel with their own vehicle, must travel *na perekladnikh*, changing their tarantass at every stage, and adding considerably to the delays which are already vexatious enough. It is also possible at times to travel in a carriage that the owner wants returned to his residence.

Travellers are served with horses in the order of their arrival, but methods are adopted to discourage racing on the road, so as to pass other travellers who are farther ahead on the same stage. The mail service, which carries no passengers, takes priority of private travellers, and it is not uncommon for a traveller who has seen his fresh horses harnessed to his tarantass to have them removed and transferred to the imperial mail which has arrived. A further necessary delay occurs at intervals when the wheels are taken off to be greased. The horses and drivers do not go beyond their next posting stage, though sometimes by arrangement a returning team and driver will exchange places with those who meet them. It is difficult to arrange to stop between the posting-stages.

The Podorozhna.—Formerly it was necessary to have a *podorozhna*, or Government permit to travel, in order to avail oneself of the advantages of the posting system, but this requirement is now obsolete, though the *podorozhna* still confers advantages. It is especially important in districts where there are many political exiles and where the restrictions on travelling are numerous, and even elsewhere a driver may refuse to supply the horses, dogs, or reindeer required. Travellers who refer to the needlessness of it are thinking of the better known parts, where its claims to give priority of treatment are somewhat in dispute.

Posthouses.—Posthouses are all on the same model. Like the other houses of the villages they are of logs. The only distinguishing mark is two wooden pillars painted black and white, surmounted by the imperial arms in front of the entrance. They differ in cleanliness and comfort, but not in architectural plan. In all there are at least two rooms, one for the postmaster and his family, the other for the travellers. In a few there may be a cot or two, but ordinarily there is merely space to lie on the floor, the traveller bringing with him such bedding or rugs as he chooses. The rooms are about 20 ft. by 18 ft., and are heated by a huge brick stove in the party-wall. The floor has no carpet, and the only furniture as a rule is a small table and two hard wooden chairs, with sometimes a wooden sofa or bench along the walls. Those near the towns are the dirtiest, and travellers will find themselves almost everywhere much troubled by insects and sometimes rats. The best are on some of the side routes ; thus on the Kupetski track, which leads south of Lake Baikal to Kyakhta, the posthouses are much better than on the Great Post Road. On the walls is a regulation price list of food and drink, but in most cases this document merely deals with the hypothesis of what would be charged if the food were there. Normally all that can be obtained is hot water, salted fish, and black bread, though in western Siberia milk and eggs are fairly plentiful. Meat is rare, and seldom good. Travellers should bring their own food, as well as their bedding and a supply of rope and a hatchet for repairs on the road. Each posting-station is provided with a black book for complaints. A record is also kept of the time of arrival and departure of each visitor, if there is a clock on the spot. Even in hotels in the towns the washing accommodation is of the most limited and primitive sort.

Yamshchiks.—The *yamshchik* is changed at every stage with the horses. He expects a gratuity (called *na chai*, ' for tea '), which is sometimes 10 kopeks, but more usually 15 from prudent travellers, for, although there is supposed to be a regulation speed of 12 versts an hour, the pace to some extent depends on the generosity of the driver's employer. An ordinary stage of about 16 miles will be done in a little over two hours, and for a journey day and night 200 versts is considered good. Three hundred versts can be secured by a special effort, but is regarded as cruel to the horses. The driver is provided with a whip, and a curry-comb of a primitive kind attached to the handle of his whip for removing ice from the horses' coats. Care should be taken that the *yamshchik* conveys the traveller for the whole distance that he has covenanted, and does not deposit him at an intermediate stage.

Horses.—A team of horses is called a *troika*, being normally three; but often a larger number are employed, sometimes as many as seven. The middle horse goes under the big wooden bow (*duga*), which is often elaborately carved, and to it are attached bells, which are only allowed to be rung on the high road, and removed on entering a town. The middle horse trots in the shafts, while the others gallop with their heads turned far out. Sometimes where the roads are bad, the horses are harnessed tandem. The horses are poor in appearance, but splendid for going: they are usually 14 to 16 hands. Every horse is given six hours' rest at the end of a stage.

Vehicles.—The ordinary conveyance is a *tarantass*, a rude, strong carriage of four wheels without springs, suited to its purpose of transit over these rough and jolting roads. The body of the carriage is borne on two long, elastic poles, which rest on the axles of the front and back wheels. In front is a box for the driver. In the carriage is no seat, but passengers, of whom there is room for two, lie on the floor, which is covered with straw, which travellers will supplement with pillows and mattresses for night travelling. Behind is a sort of hood, and the whole back part of the carriage can be cut off entirely by stretching a tarpaulin to the coach box when it is wet. The luggage can be strapped behind. The shafts are made fast to the *duga* to keep them clear of the horses' sides. Two-wheeled cars are called *provoloki*, and are especially used in the taiga. A more primitive and rougher kind of conveyance is the *telega*, which is often used as a cart to carry luggage, and as such often accompanies the convoys of prisoners who march to their distant settlements in Siberia. In towns there are other kinds of vehicles: e. g. in Tomsk the droshki is a low jaunting-car, in which two travellers sit back to back on a plank about 18 inches wide; in Irkutsk the droshki has been compared to a sort of hoodless bath-chair.

Condition of Roads.—The regulation breadth of the main roads in the western parts is 21 ft., and of the smaller roads 14 ft.: the great breadth of the road is due to the cheapness of the land. At the side are broad stretches of grass for riding and walking, and even for driving, when that becomes impossible on the road. Also the telegraph line has a broad space where all trees and bushes are removed, and there also it is possible to ride and drive, when necessary. Even bigger departures from the road are not unusual. One traveller mentions that east of Tomsk there is practically no road, but the *yamshchiks* take their own route. Another traveller records that it is not uncommon in the part west of Irkutsk to leave the road, and take a short cut through the taiga by a track. The methods of road-making are primitive. Apparently the best surfaces are in the Government of Yeniseisk. In the Governments of Tobolsk and Tomsk the roads are very muddy; east of Irkutsk they deteriorate, and beyond Lake Baikal they are the worst of all. The centre of the road is often cut to pieces by the big caravans that pass over it, especially in early autumn, to such an extent that traffic is almost impossible. The best part of the *Trakt* is between Krasnoyarsk and Irkutsk; near Kansk it has been declared excellent, with a hard even surface. Verdicts pronounced on special parts of the road differ exceedingly; near Abatskoe, where the roads from Tyumen and Tobolsk meet, a motor-driver has pronounced it a sea of mud, stretching wide on both sides of the road, and only passable by the indication of the telegraph posts, and between that point and Omsk the ruts were said to be sometimes 20 inches deep. The ruts are so marked in many roads that *yamshchiks* will sometimes refuse to accept vehicles of greater width. In the vicinity of many towns the roads are bad; specially unfavourable comments have been made on them at Tomsk, near Bogotolskoe (at the boundary of western and eastern Siberia), Achinsk, west of Krasnoyarsk, and west of Chita, where a somewhat sandy track is described as winding capriciously through the depths of a thick forest. The road from Irkutsk to Lake Baikal at Listvenichnoe is described as good, and the road on the south-east side of the lake as difficult but good. A particular portion of the road will change its character very rapidly, and different accounts of its surface are given by different travellers who have passed over it at comparatively short intervals of time.

Streets in Towns.—In the towns conditions are at their worst: there is no regular paving. Nansen describes the streets of Yeniseisk as ' muddy and full of deep ruts '. Along the sides of the streets are wooden planks to serve as footways: usually these are fringed with a line of posts to mark them off from vehicular traffic, when roadway and footpath alike are under snow.

Theft.—Thieves, who are often escaped convicts (*brodyagi*), are alleged to frequent certain parts of the roads; parts that have been specially indicated are between the Lena and Irkutsk (1904), and the parts of the *Trakt* between Nizhne-Udinsk and Kutulikskoe (1889). The *brodyagi* were often in alliance with the *yamschiks*; they were usually armed with bludgeons, but did not carry firearms.

Verst-posts, telegraphs, &c.—Along the sides of some of the roads are verst-posts with square tops cut in such a way that the traveller can easily see the distance that he has traversed from the last posting-station and the distance that he has to pass before reaching the next. At the posting-stations are boards affixed which show the distances to Petrograd and other important towns. There are telegraph stations along the important roads, the lines of telegraph being usually in broad lanes at the side of the road. It has been recorded by one writer that in some places over the steppes the telegraph wires are laid along the ground for 20 or 30 versts in order to avoid the violent storms that sweep over those localities. This is said to be especially true of the neighbourhood of Mariinsk and Krasnoyarsk. At the entrance to villages are gates, guarded by a watchman: these are kept closed during the summer, to prevent animals straying into the village, but are open during the winter.

Bridges and ferries.—Bridges are flimsy and short-lived wooden structures, with low hand-rails. For all the wider rivers there are ferries. On the Yenisei near Krasnoyarsk the river is crossed where two islands lie in mid-stream. The islands are reached from the banks by ferries, and are united by a short bridge. Among other places on the great post road there are ferries at Omsk, Dubrovinskoe, Tomsk, Mariinsk, Achinsk, Bolshe-Kemchug, Kansk, Nizhne-Udinsk, and Ziminskoe. There are four kinds of ferries—(1) one kind is propelled by horses, who work it by being driven round in circles —there is an example of this found at Tomsk; (2) a second kind is propelled by oars; (3) a third kind is a pendulum-boat, which takes advantage of the current of the stream—there is an example on the Yenisei near Krasnoyarsk; (4) and a fourth is a cable-boat worked by a wheel.

Sledging

The normal method of travelling throughout Siberia during the winter and in northern Siberia during the whole year is by sledging. The sledges are drawn by horses, reindeer, or dogs in the various districts, horses being employed in the south, and reindeer and dogs in the north, the latter exclusively by certain tribes such as the Kamchadal.

Nature of ordinary sledge.—The ordinary sledge is called a *narta*: it is a narrow vehicle from 9 to 14 ft. long by 30 ins. broad, fitted with a movable hood (*koshma*), which can be drawn completely over during storms or severe cold: the traveller must beware of letting it rest on his face during sleep, for it may cause frost-bite. The *narta* is a very light vehicle, and pitches heavily when it is traversing rough ground. The traveller can lie at full length. The runners are usually made of birch poles, and the fabric is kept together with cords, as nails would be jerked out almost at once. A Samoyede sledge has two large thick runners curved up at the end in front to a height of two feet. On each side are four uprights, placed rather close together towards the rear. These slope upwards and inwards until at 2 ft. they are united together by stout bars, which act as cross-overs and make the floor of the sledge: the long pieces are called *bereznyas*; on this floor is put the luggage, and the driver sits on it or just in front. In some parts of Siberia the driver sits beside a perpendicular ' bow ' of stout wood which rises some 4½ ft. from the ground, about one-third of the distance from the front to the back of the sledge. He avoids obstacles by pulling this way or that by means of the ' bow '. The *harray* is a stout steel-shod stick 15 ft. in length with a cord attached to the end ; it can be used as a brake by putting it between the runners, or the sledge can be anchored to it. The point can be used for testing the ice when crossing a river.

Other kinds of sledges.—There are also the *balog*, a sort of family sledge or gipsy-car on runners, covered with tarpaulin and skins, which can even contain a cooking-stove; the *vashok*, which is entirely enclosed ' like a huge brougham on runners ' ; the *kachovka*, a great open sledge, roughly made of wood and matting, and with no covering save a piece of matting or felt to spread over the recumbent travellers ; the *pavoska*, which is described as a large, deep, roughly-built sledge, open in front, but covered in at the back with a canvas hood lined with thick felt : the driver's seat consists of a flat board, from which slope outwards and downwards a pair of stout poles to save the vehicle in case of collision.

Reindeer harnessed to sledges.—Between two and five reindeer are normally used for the sledge, though as many as eight were seen by Nansen drawing a *balog*. The Lapps, who have bigger reindeer, only use one at a time. A reindeer will draw a load of 400 lb. over snow, and to a sledge which had a load of 800 lb. of blubber only two were harnessed. The Samoyede harnesses his reindeer by an ingenious system (described at length in Jackson, *Great Frozen Land*, p. 115), which compels each deer to do his share of the work, by a couple of *chulki* (' tackle-blocks ') made of wood or walrus ivory, through which the trace runs from the near to the offside reindeer. The same writer drove three reindeer for 120 versts (1 Samoyede or Reindeer verst = 4 Russian versts) within 12 hours without feeding them, and they went the last 10 versts as well as the first, and were quite fresh after two days' rest. The reindeer is independent of roads ; he will find his food under the snow, which he scrapes away with his hoofs for a depth of 2 ft. or more, but ice may cut him off from his moss, and then he will soon be in poor condition. All that he requires is protection from wild beasts, among his special enemies being the wolverine, black bear, and polar wolf. If unharnessed at any time he must be hobbled, or he will return even 40 or 50 miles to rejoin the herd. His most frequent complaint is hoof-swelling. (For the breeding and habits of the reindeer see Chapter V.)

Reindeer riding.—The reindeer is sometimes ridden, but not on the back, which would cripple it. The saddle is on the shoulders, and the rider gains his seat by means of a pole in his right hand, not touching the saddle with his hands. To keep his seat he practises a swinging movement, balancing himself with his pole, but it is unwise to put it to the ground to steady himself, for he will probably be dismounted. Again, if he grips with his knees so that the cushion slips back, the reindeer, feeling the weight, will bend under his haunches and deposit the rider on the ground. The Soyots use reindeer for riding among the Sayansk Mountains, as their food is more easily available than that of horses. They ride with two reindeer, one serving as pack-animal and relief. A Soyot, whose average weight is 4 puds in his furs, can ride a reindeer, while a Russian cannot.

Dogs harnessed to sledges.—The dogs employed by the Samoyedes are like Eskimo dogs, but somewhat smaller; those used by the Kamchadals are said to resemble wolves. Usually there are 6 or 8 dogs in a team. Most tribes are kind to their dogs, and in some places (as along the Yenisei) they are even given the warmest place over the brick-oven to sleep on when they come in tired, and whoever may be there has to vacate it for them. They can go 60 miles at a stretch without being fed. In the Yenisei district a good sledge dog is worth 160 roubles, but such are not easily obtained, for a dog is of very little use for draught purposes unless he has been reared by his owner. A young puppy can be bought for a rouble. Older dogs are seldom bought except for breeding purposes. When running in the sledge dogs, who have known one another since they were puppies must be paired : it is the only way of securing peace ; but dog teams will fight one another when they get the opportunity. Ordinarily a dog country is apart from a reindeer country ; but if the dogs get scent of a deer, they will become unmanageable. When dogs begin to paw the snow it is said to be a sign of a coming storm. The team dogs of eastern Siberia serve till they are 10 or 12 years old, but begin to deteriorate after 6 or 7 years. They suffer frequently from rabies in the spring, but sometimes continue to be harnessed, though muzzled, when rabid. They also suffer from palsy and cramp : their feet are apt to become sore and are often made to bleed. Sometimes then they are put into shoes, but they greatly resent this treatment. They can draw considerable weights. Haviland describes a pack in which each sledge with eight dogs could draw 1,440 lb. of goods and two men, and the leader by himself could draw 180 lb. on the sledge. On a good surface they could travel as much as 66 miles in 3 hours. The Russian, Koryak, and Kamchadal dogs feed exclusively on fish, the Chukchee dogs on intestines of seal and the blubber of seals, whales, and walrus.

TELEGRAPHS AND CABLES

Siberia

The telegraph system has been extended in Siberia in recent years, but there is still no close network of wires. Away from the railways and the chief rivers few places are on the system, and the north is almost devoid of telegraphic facilities.

The trunk system consists of a line more or less following the railway between west and east and linking the telegraphic system of European Russia with Vladivostok, and via the Chinese Eastern Railway with Kharbin (Harbin), Mukden and the Chinese telegraphs. Most telegrams between Europe and Siberia pass either by Yekaterinburg or by Chelyabinsk, but the South Siberian Railway, when built, will afford a route via Orenburg. Russian Central Asia has its own lines of telegraphic connexion with Europe.

From this trunk line, which traverses the most populated regions of Siberia, several branches run to north and south, but except in western Siberia there are no alternative routes to the main line, within Russian territory. The Amur telegraph lines through the Amur and Ussuri districts form an all-Russian route alternative to the more direct route through Manchuria along the Chinese Eastern Railway.

The four chief lines to the north more or less follow the four great rivers. From Omsk a line runs north along the Irtish to Tara, across country to Tobolsk to avoid the swamps of the lower Irtish and thence down the Ob to Kondinskoe and Berezov. From Tobolsk a branch leaves this Irtish–Ob line and goes south-west to Tyumen where it joins the trunk line along the Siberian Railway. From Krasnoyarsk a line runs north to Kozachinskoe and follows the Yenisei to Yeniseisk and Turukhansk. On both this and the Ob line telegraph stations are widely separated in the north. From Irkutsk a line runs north-east to the Lena at Manzurskaya and then follows the Lena to Yakutsk with branches to Bodaibo and to Vilyuisk. From Yakutsk a line goes across country, following the rough track to Okhotsk. This is the most northerly line in Siberia.

In Kamchatka there is a line between Tigilski and Petropavlovsk, but it is said not to be in working order. Lastly at Chita the Amur line branches from the old trunk line which runs through Manchuria to Vladivostok along the railway. The Amur line leaves the railway at Stryetensk and keeps near the Shilka and the Amur to Khabarovsk where it turns south up the Ussuri to Vladivostok. A branch line from Khabarovsk follows the Amur to Nikolaevsk, sending branches across the northern end of the Sikhota Alin Range to de Castries Bay and to Cape Lazarev, whence cables cross to Sakhalin. On Sakhalin there is a line from Cape Pogobi to Alexandrovsk and Dué. The Amur line continues eastward from Nikolaevsk to Chnuirrakh Point at the Amur mouth.

The southern branches from the west and east trunk telegraph line are more numerous. Most westerly in Siberia is a branch from Petropavlovsk to Atbasar and Akmolinsk in the steppes. From Omsk a very important line runs more or less along the Irtish to Semipalatinsk, whence it continues southward to Sergiopol, Kopal, Vyerni, and Tashkent in Russian Central Asia, connecting with the lines to Krasnovodsk on the Caspian Sea and to Orenburg in European Russia. This southern line sends several branches towards Mongolia, to Zaisan, Chuguchak, Kuldzha and elsewhere. At Chuguchak (Tahcheng or Tarbagatai) and Kuldzha there is connexion with the Chinese telegraph system.

Another branch leaves the trunk line at Novo-Nikolaevsk and goes by Barnaul and Biisk to Kosh-Agach on the Mongolian frontier. From there to Kobdo, where the Mongolian telegraph system begins, telegrams are carried by Cossack post. In 1913 Russia was granted a concession by Mongolia to extend the telegraph line to Kobdo. A fourth important branch leaves the trunk line at Achinsk and goes to Minusinsk and Grigorevski on the frontier. In the Baikal and Trans-baikal regions there are several short lines southward to places on the frontier, including those from Kultuk to Tunkinsk, Verkhne-Udinsk to Troitskosavsk (meeting the Chinese telegraphs at Kyakhta), and Chita to Mangut. Finally there are (*a*) the coast line northward from Vladivostok to Tyutikha Bay, which is being continued northward in order eventually to meet the Amur telegraph line ; and (*b*) the line southward from Vladivostok to Novo-Kievskoe near the frontier of Korea.

Considerable difficulty is experienced in erecting telegraph wires in many parts of Siberia on account of the thick forests, swamps and other obstacles, but the difficulty of upkeep is still greater. Trees fall and interrupt communication, and natives are continually stealing the wire. Siberian telegraphs seldom work satisfactorily, and there is generally delay, not always unavoidable, in the transmission of messages.

The only submarine cables touching Siberia are in the far east. Two cables, belonging to the Great Northern Cable Company of Copenhagen, connect Vladivostok with Nagasaki in Japan ; and there are Russian cables from Lazarev and de Castries Bay across the Straits of Tartary to Cape Pogobi and to Alexandrovsk, respectively, in Sakhalin. A Japanese cable also runs between Alexandrovsk, Todo Shima, a Japanese island off southern Sakhalin, and Hokushu.

There are no cables to Kamchatka or across Bering Strait to America.

Arctic Russia

In Arctic Russia, except around the White Sea and on the Murman coast, the system of telegraph lines is little better developed than in Arctic Siberia

Murmansk and other ports on the Murman coast are linked with the Russian telegraph system both along the route of the Murman Railway and round the coast of the Kola Peninsula. All White Sea ports and most places on the Northern Dvina also have telegraphic connexion, but in the Pechora region Ust-Tsilma is the only place with a telegraph line with the exception of a few stations in the extreme upper reaches. There is no telegraph line to Siberia north of the railway.

There is no permanent through telegraph line to Finland north of Lake Ladoga, but a Russian military line runs from Rovaniemi to Pechenga. The Russian and Norwegian systems meet on the Voriema River, but are not connected.

There are submarine cables from Alexandrovsk to Peterhead, Scotland, and from Alexandrovsk to Arkhangel.

Wireless Telegraphy

In recent years several wireless stations have been built in the far north and in the north-east, and others are projected.

On the Murman coast there are stations on the Ribachi Peninsula, at Alexandrovsk, Murmansk, Teriberski, and Svyatoi Nos ; on the White Sea at Kandalakska, Kem, Solovetski, Arkhangel, Sosnovets Island, Cape Voronov and Kanin Nos. The nearest Norwegian stations are at Ingö, near the North Cape, and one at Green Harbour, Spitsbergen, each with a normal range of 480 miles. To facilitate the navigation of the Kara Sea three stations have been erected, on Yugor Strait, on Vaigach, and at Cape Mare-Sale on Yamal respectively, each with a normal range of 150 nautical miles. Others in Novaya Zemlya and elsewhere are projected (see Vol. II, Chap. V). A station at Obdorsk on the Ob delta is contemplated, and one at Dickson Island at the mouth of the Yenisei (range 1,700 miles) has been in operation for several years.

In the far east there is a station at Okhotsk (range 130 miles) which communicates with one at Nayakhanskoe on Gizhiga Bay (range 130 miles), one at Novo-Mariinsk at the mouth of the Anadir (range 130 miles), and another at Markovo farther up the Anadir. Stations are reported to be under construction at Yamsk on the Sea of Okhotsk, and at Sredne-Kolimsk on the Kolima.

On the Amur there are wireless stations at Khabarovsk (military), at Nizhne-Tambovskoe (military) and at Nikolaevsk at the mouth (range 240 miles). The Nikolaevsk station communicates with one at Petropavlovsk in Kamchatka (range 240 miles). A second station is contemplated in Kamchatka at Tigilski on the west coast. Stations are proposed on Bering Island and on one of the Commander Islands. The station at Nikolaevsk also communicates with one at Kerbinski on the Amgun (range 170 miles). On the Ussuri there is a military wireless station at Iman.

At Vladivostok there are three wireless stations, one belonging to the army and two to the navy. Japan has several stations within range of Vladivostok. A number of the Russian vessels which ply in far eastern waters as well as the Russian ice-breakers are fitted with wireless installations.

There are no Russian wireless stations on Bering Strait, but stations are proposed at Providence Bay and Cape Dezhneva. The nearest United States stations of long range are those at Fort St. Michael and at Nome on Norton Sound in Alaska, both controlled by the United States Army, and those at Unalaska in the Aleutian Islands and at St. Paul in the Pribilov Islands, both operated by the United States Navy. The station at Nome communicates with the one at Novo-Mariinsk.

CHAPTER XVII

RAILWAYS

General Considerations—The Siberian Railway—The Amur Railway—The Ussuri Railway—The Altai Railway—Other Siberian Railways—The Murman Railway—The Arkhangel-Vologda Railway—New and Projected Railways.

General Considerations

Progress in Construction

The total mileage of Russian railways open to traffic (47,480 miles in 1914) is large in comparison with that of other countries, but small in proportion to the area of the Empire. Of this total about 8,000 miles are in Siberia. Railway construction in the Russian Empire was proceeding, before the war, according to a definite plan. For the years 1914–19, a programme was drawn up for constructing about 2,330 miles of line each year, but the Commission under General Petrov had been of opinion that something more like 4,000 miles each year were necessary for the needs of the Empire.

The Russians build railways very quickly, so that they have been able after the outbreak of a war to make radical and timely improvements in their system of communications. In the course of the war with Japan they were able to finish the Circumbaikal section of the Siberian Railway in January 1905, a work of immense difficulty, as the section contains 40 tunnels. Meanwhile a line had been laid across the ice on Lake Baikal, from Baikal to Tankhoi, a distance of 25 miles. The work was accomplished between February 9 and March 1, 1904, at a laying rate of about 1⅓ mile a day. From the outbreak of war in 1914 to the spring of 1916 a thousand miles of the Murman Railway were built, and large sections necessary to complete the Amur Railway were constructed.

Military Importance

In the Russian Empire railways are particularly important from a military point of view, owing to the vast distances between place and place, and owing to the state of the roads. Considerable advance in road-making took place in the first thirty years of the nineteenth century, but with the advent of railways, less attention has been paid to roads. There are very few roads in the Russian Empire which can support artillery, or which could allow a railway-line to be laid along their surface. The rivers, though suitable for transport, suffer from many disadvantages; the Volga, for instance, runs into the Caspian, which is a closed sea; the great Siberian rivers are all frozen in winter. Nevertheless, the rivers and canals of Russia have great possibilities in the development of military transport.

Characteristics

The prominent characteristics of Russian railways are, first, that they are cheap to build, costing about £10,500 per mile in Europe, a little over £5,000 per mile in Asia. This cheapness is partly due to the general flatness of the country, and the straightness with which the line is traced. A second characteristic is that tunnels are few; they are avoided, for instance through most of the course of the hilly Amur Railway, by steep gradients and sharp curves. A third characteristic has been already mentioned, the straight tracing of the line, so that towns are left on either side. The straightness of the track makes the measurement of distances off a map more approximately accurate than might otherwise be expected. Fourthly, most of the lines are single, but with embankment or bed provided for a double track.

Gauge

The standard Russian gauge is 1·524 metre (5 feet); there is a large number of small branch and feeding lines (in

European Russia) which have a narrow gauge, varying from 1·006 metre to 0·75 metre (3 ft. 3½ in. to 2 ft. 5½ in.). The wide gauge admits of heavy loading of trucks, but this advantage cannot be much used as the rails are not sufficiently heavy, nor the road-bed good enough, to take very heavy trains. But great improvements have been effected in recent years, especially on the Siberian Railway.

The lessons of the Franco-Prussian War gave a great impetus to strategical railway building in Russia. The broad gauge was adopted, it is said on the advice of an American engineer, Major Whistler, to prevent the German and Austrian rolling-stock in time of war being put on to the Russian lines. The German and Austrian gauge is 4 ft. 8½ in. The Japanese gauge is 3 ft. 6 in., but 4 ft. 8½ in. was adopted as the standard in 1916. In the Russo-Japanese war, after the battle of Mukden, the Japanese narrowed the gauge on the Southern Manchurian Railway, but it took them 39 days to do it for 34 miles. It has since been widened, see p. 342. The Germans, however, claim to have means for dealing rapidly with the task of converting the Russian gauge. Owing to the width of the embankment it is obviously easier to convert a broad gauge to a narrow one, than vice versa.

Permanent Way

In Russia in Europe, the sleepers are said to be sunk in the permanent way, so that the track can be used for marching. The surface of the bridges, which is open, would have to be covered, if the tracks were to be utilized for this purpose. This practice of sinking the sleepers seems to have been followed in a large number, if not in all the sections of the Siberian Railway. The railway tracks in Russia are often the only unflooded roads to be found in a large district.

Rolling-Stock

The amount of rolling-stock is not so great as in Germany or Austria, and owing to the great distances it is difficult to collect much together for a particular purpose at any one place. In 1911 there were 21,121 locomotives, burning either coal, oil, or wood; 24,487 passenger coaches; 469,063 goods wagons. But all these were said to be rather old. It is difficult to add new rolling-stock rapidly from Russian work-shops; construction is retarded by the necessity of building different types of engines, some to use coal, others wood, others oil. In 1910 the Russian locomotives were stated to be 30 per cent. less powerful than those of Germany. The average speed is 13 miles per hour, on fast trains, 33 miles. The want of sufficient rolling-stock was still noticeable in 1916. But locomotives and cars were imported in parts from the United States to Vladivostok, and were erected in shops created during the war at Pervaya-Ryeka, on the main line five miles outside the city. In spite of the marshy, low-lying nature of the country, which necessitated much filling-in, a large system of sidings and workshops was completed by the beginning of 1916. At first only 5 or 6 cars were erected each day, but the output was expected within a short time to reach 100 to 150 a day.

Loading Capacity

The broad gauge admits wagons with a large carrying capacity. Ordinarily a wagon can take 32 to 40 men, or 6 to 8 horses. Most cars for carrying men are heated with a stove. A military train has from 30 to 50 wagons, 35 to 40 being the commonest number. In its composition, a field-kitchen is also frequently included. The American cars purchased in 1915–16 were of the bogie type, 42 ft. long, and could carry 2,400 poods, i. e. nearly 39 tons. 14,000 of these cars were under order from America at the beginning of 1916, along with 450 American locomotives. As well as the newly constructed shops at Pervaya Ryeka, the machine-shops of the Chinese Eastern Railway at Kharbin were much used, being able to erect 4 or 5 locomotives per diem.

Stations

Most of the Russian stations are some distance, varying from ½ mile to 10 miles, from the town or village after which they are named. On the Chinese Eastern Railway the distances are sometimes even greater. The stations are divided into classes from I to V. Class I is designed for changing locomotives and personnel of passenger trains; Class II for dealing in a similar way with goods trains. Class III is designed for maintaining locomotives and personnel for local work. Class IV is designed for occupation by traffic staff only. Class V exists for the same objects as Class IV, but on a smaller scale. On the Siberian line at any rate the objects of the classes have been to a certain extent confused; thus Classes I–III have all engine depôts, and have been classified only according to size. All Classes IV and V have arrangements for watering locomotives, as well as living accommodation for traffic staff. A 'crossing' is a point on the line, with (on the Siberian line) two loops in addition to the through track, and with sidings. There is a pointsman and accommodation for him at each crossing. When a line is doubled, the crossings may be eliminated. The stations of Russian railways generally have low broad platforms. In addition, being generally some way distant from their town or village, many stations have wide open spaces near by, very suitable for parading troops prior to entraining or on detrainment. On the Siberian Railway, the stations of Kharbin, Irkutsk, and Baikal have not these open spaces. For off-loading trains, ramps have to be used. Lifting appliances, cranes, &c., are not commonly used or provided. Most of the larger stations, even down to Class IV, have good buildings, many of stone, which are constructed on a generous scale.

Military Personnel and Control

The military personnel dealing with railways consists of 17 battalions, comprising 77 companies. There is a field-railway park with 100 kilometres (62 miles) of line, and there are field-railway depôts. The lines are divided into districts, allotted to the transport line-commandants and station-commandants. In war, all the lines, at least in the theatre of operations, go under military control. A railway-battalion consists of two companies of constructors, two of 'exploitation' employés, and two companies of reserve. There are three railway-battalions in the engineers of the reserve.

Government and Private Enterprise

About 69 per cent. of Russian railways are State-owned. All the Siberian Railway to the station of Manchuria, was built and is worked by the Government; this is true also of the Amur Railway, which runs entirely through Russian territory. Several new branch lines, the Slavgorod Railway, the Altai Railway, and the Minusinsk Railway are built and worked by private enterprise. The Chinese Eastern Railway is under a private company, rather closely related to the Government. The private lines are said to be better constructed than the lines built by the State.

Civil Control

Three official bodies deal with the railways: (1) the Ministry of Ways and Communications, which deals with the technical and administrative mechanism of the railways, including construction; (2) the Ministry of Finance, which deals with the raising of capital, with tariffs, and kindred matters; (3) the Control of the Empire, which inspects receipts and expenses, and makes up the complicated accounts between them. These bodies regulate private as well as State lines.

Defects in Management

Circumstances have prejudiced the successful working of the Siberian Railway system. The line is made up of separate railroads, each having its own independent administration and its head-quarters in Petrograd. The control over these railroads has not been properly co-ordinated; overlapping is constantly occurring; the different managements have never tried to work with one another for the improvement of the whole system, but have merely looked after their own sections. The initiative of local officials has been hampered by circular instructions sent out by the Ministry of Ways and Communications. The system for watering engines is cumbrous and involves long delays.

The Siberian Railway

History

The idea of the great Siberian Railway is said to have been originated by Count Muraviev-Amurski, who became Governor of Eastern Siberia in 1848. He founded Nikolaevsk at the mouth of the Amur in 1850. His idea was to connect the Maritime Province with Russia; when the construction of the Siberian Railway was actually taken in hand in 1891, it was intended to carry out Muraviev's idea by building the line all the way through Russian territory, avoiding Manchuria. The earliest definite proposal was that of an English engineer called Dall, who put forward a plan for a horse-railway, to be laid from Nizhne-Novgorod by way of Kazan and Perm, to a Siberian port on the Pacific. After 1870 the railways of Russia took a steadily eastward trend. In 1877 the line to Orenburg was opened, and in 1878 the Ural Railway, ending at Tyumen. By the year 1890, there were three lines which had claims to be used for the great extension on to Vladivostok: one was the Ural Railway, a very important one on account of the mines and iron-works adjacent to it. The second was the Samara–Zlatoust Railway ending at Miyas. The third was the Samara–Orenburg line. The Samara–Zlatoust line was chosen as it offered the shortest route through Nizhne-Udinsk, and as it ran through the fertile black-earth region of western Siberia. The building of the line was begun in 1891, in different sections, and completed during the Russo-Japanese War, in 1905. Only Russian labour and Russian material were used, except perhaps on the Circumbaikal section which was finished in a hurry during the Russo-Japanese War. At various points along the route, steamer-wharves were made on the rivers, saw-mills and foundries were opened, and these still exist and offer means for railway-construction.

Chinese Eastern Railway

The original intention was that the Siberian Railway should run from Samara to Vladivostok by way of Stryetensk and Khabarovsk. But as the route from Stryetensk to Khabarovsk (now called the Amur Railway) offered technical difficulties, an alternative route was adopted, through Manchuria, which was Chinese territory. In 1896 an agreement was signed between the Chinese Government and the Russo-Chinese Bank, for the formation of the Company of the Chinese Eastern Railway. This railway was to be built to connect the Transbaikal Railway with the South Ussuri Railway. The Company was given the power to work coal pits and to engage in other industries ' in China '. The shareholders can only be Russian or Chinese subjects. The Russian Government undertook certain financial guarantees with regard to the ' obligations ' of the Chinese Eastern Railway, but not with regard to its share-capital. The Company was to be under a Directorate, which was to sit in Peking and Petrograd. There were to be a chairman and nine directors, the chairman being named by the Chinese Government, the directors to be chosen by the shareholders. The line is to be leased to the Chinese Eastern Railway Company for 80 years, and at the end of this period is to pass, without any payment, into the possession of the Chinese Government. The luggage of passengers, and all goods, in passage from one Russian station to another over the Chinese Eastern Railway, are free from Chinese customs. In the tracing of the line, villages, towns, and cemeteries were to be left aside. The gauge of the Chinese Eastern Railway was to be the same as the Russian (5 ft.). After the Russian Government obtained a lease of the Laio-tung Peninsula in 1898, the Chinese Eastern Railway was extended from Kharbin to Dairen and Port Arthur (614 miles). By the Treaty of Portsmouth, August 1905, Japan took over the southern section of this line, from Port Arthur to Kuan-cheng-tsu (467 miles). From Kharbin to Kuan-cheng-tsu it remains under the Chinese Eastern Railway Company. The line from Kuan-cheng-tsu to Port Arthur is now called the South Manchurian Railway, and since it was taken over by the Japanese has been converted from a 3 ft. 6 in. to a 4 ft. 8½ in. gauge. In practice the goods are transferred at Changchun, 1 mile south of Kuan-cheng-tsu. The Chinese Eastern Railway is really under the Russian Ministry of Finance, which controls it through the Russo-Chinese Bank.

Amur Railway

The disadvantage of the Manchurian route, from the Transbaikal Railway to Vladivostok, is that as far as Pogranichnaya, 144 miles from Vladivostok, it does not run through Russian territory. It is for this reason that the Amur Railway, connecting the Transbaikal line with Khabarovsk, and so by the existing north Ussuri Railway, with Vladivostok, is so important. It runs all the way through Russian territory, has embankment for a double track, and will probably in time become the main artery of communication with the old Siberian Railway. It runs some way north of the river Amur, instead of along its valley, so as not to be too near the frontier in case of war (see p. 347).

Capacity

The Siberian Railway was rather hurriedly and cheaply built. The ballast was thin, and the rails were only 54 lb. to the yard. They were afterwards replaced by 72 lb. rails. A report of October 1915 states that the line has been relaid with a still heavier rail, apparently 80 lb. per yard. The wooden bridges are also said to have been replaced by bridges of steel and masonry. While the Siberian and Transbaikal Railways were only single-track, the number of trains that could pass each way in a day was 10½ (21 in all on the line) on the Siberian Railway, 7½ on the Transbaikal. During the Russo-Japanese War the Transbaikal Railway is said to have been worked to take 12 trains a day each way.

Effect of Double-tracking

The doubling of the track between Omsk and Karimskaya (with the exception of the part round Lake Baikal) and the strengthening of the line in the mountain sections permit 34 trains now to proceed in either direction. The through lines at the stations can take 60 trucks. The bridges were built for a single line, and many of them have yet to be adapted for a double line, even in the part where the line is already doubled.

General Description

The name Siberian was originally applied to the railway from Samara to Irkutsk. From Irkutsk to Manchuria was known as the Transbaikal Railway, in the days when Lake Baikal formed a huge gap in the line, traversed by a train-ferry from Baikal Station to Misovaya. To this was added the Circumbaikal Railway, completed during the Russo-Japanese War in 1905. On the Siberian Railway the old line from Yekaterinburg joins the Moscow–Samara line.

Chelyabinsk to Ob.—From Chelyabinsk to the River Ob, the line passes through a fertile, black-earth country, very good for agriculture, especially in the Ishim and the Baraba steppes. The only difficulties are the Rivers Tobol, Ishim, Irtish, and Ob. The country is studded with clumps of dwarf elms and willows, which are only good for firewood. The population is agricultural and pastoral. As far as Chulim (802 miles from Chelyabinsk) the greatest grade is 1 in 165, the curves 1,750 ft.; from Chulim to Ob the grade is 1 in 135, the curves 1,750 ft.

Ob to Irkutsk.—From the Ob to Irkutsk, the line goes through hilly country, with some large rivers, the Tom, Yaya, Kiya, Chulim, Yenisei. There are a number of smaller rivers to be crossed, and, as the valleys are fairly close together, the maximum grade is frequently reached. The country is thickly wooded, and there are very few clearings. After Achinsk the country becomes mountainous. At the higher elevations the forests consist of giant conifers. Forests are traversed for 200 miles in long stretches of 25 to 45 miles without a break.

The population consists mainly of settlers along the postal road from Tomsk to Irkŭtsk. A great deal of earthwork had to be used in the construction of this part of the railway, and in many places the embankment reaches 19 or 20 ft. in height. From Ob to Achinsk the grades are 1 in 111 and 1 in 125, the minimum curve 1,750 ft. From Achinsk to Kansk, the grade is 1 in 66·6, minimum curve 1,050 ft.; from Kansk to Nizhne-Udinsk, 1 in 57, and 1,050 ft.; from Nizhne-Udinsk to Zima, 1 in 91, and 1,050 ft.; from Zima to Polovinnaya, 1 in 57, and 1,050 ft.; from Polovinnaya to Irkutsk, 1 in 111, and 1,050 ft. Since these figures were obtained in 1908, the track has been doubled from Omsk to Karimskaya, and opportunity was probably taken to ease the grades and to widen the curves in many places. Indeed, the Transbaikal sections, which in 1907 were stated to be the weakest links of the railway, are now, since the doubling of most of the track, spoken of as the best portion of the whole line. From Mariinsk to Irkutsk, the railway follows the Great Siberian Post Road more or less closely. Only between Krasnoyarsk and Kansk is it distant between 18 and 19 miles, owing to topographical difficulties.

Transbaikal.—The Transbaikal Railway presented great difficulties in construction. After leaving Irkutsk the line follows the left bank of the Angara to the shore of Lake Baikal. In this section retaining walls were necessary to protect the toe of the embankments from the action of the river, which has a speed of 7 ft. per second. The Circumbaikal section starts from Baikal station, and for 52½ miles, to Kultuk, follows the sinuous and rocky southern shore of the lake. The banks of the lake are 900 to 1,300 ft. above the water-level. It is in this section that tunnels are first found on the Siberian Railway. Between Baikal and Kultuk there are 40 tunnels, with a total length of nearly 4½ miles. Between Kultuk and Misovaya the line has to take the slopes of the Shamanski Spur (at 96½ miles from Irkutsk), the slopes of the Khamar-Dabansk (at 100⅓ miles); at 101⅔ miles it penetrates the spur of the Kerkidal. From 115 miles to 125¼ the line leaves the lake. From 173 miles it follows a plain until Cape Malinovskaya is reached at 176¼ miles. The country is almost uninhabited, and the soil nearly always frozen. Besides the tunnels, there are many cuttings in the hard rock ; the deepest cutting is said to be 74 ft. On the Circumbaikal portion of the line the grade is 1 in 125, with minimum curve of 1,050 ft.

After Misovaya the line loses its mountainous character, till it ascends the Tsagan-Da Range at 372 miles from Irkutsk. The valley of the River Khilok is well cultivated. At 609 miles the line takes the summit of the Yablonoi Range, at about 3,137 ft. above sea-level. In descending the line falls 1 in 57, with curves of 1,050 ft., for a distance of 8 miles. The valley of the Ingoda presented great difficulties, as it is winding and the mountains are steep and come so close as to leave only a small strip free, which is often overflowed. The line follows the left bank of the Ingoda, and it is only in a few places that the embankment can be said to have sufficient room. The upper Shilka valley is just as bad, with rocky slopes to be crossed, some as much as 6 miles broad.

Along the Transbaikal line there is plenty of stone and timber (Siberian larches, pine, fir). The grades and curves are as follows : from Irkutsk to Baikal, 1 in 1,075 on curves of 1,260 ft., and 1 in 143 on curves of 1,050 ft. ; from Misovaya to Karimskaya, 1 in 107·5 on straights, 1 in 143 on curves of 1,050 ft., with the following exceptions : at about 372 miles from Irkutsk, and again at about 609 miles, the grade is 1 in 57 on straights, and 1 in 66 on 1,050 ft. curves ; at Karimskaya for 13⅓ miles the grades are 1 in 66. From Karimskaya to Manchuria the grade is 1 in 106 on straights, 1 in 143 on minimum curves of 1,050 ft., except that for 19½ miles near Buryatskaya (743 miles from Irkutsk) and 5⅓ miles near Borzya (at 864¼ miles) there are grades of 1 in 57 on straights and 1 in 66 on minimum curves of 840 ft.

The watering of trains is difficult on the Transbaikal Railway. The ground is permanently frozen ; even in summer, although the soil is thawed for 3 to 10 ft., the ground beneath remains frozen to a depth of 120 ft. Water-pipes therefore cannot be buried below frost-level, in order to prevent them from freezing. On the Transbaikal Railway the suction and delivery pipes were placed in galleries, warmed by steam or hot water or hot air. Steam is only used for warming short lengths of suction-pipe. The delivery pipes, for distances up to 800 ft., are warmed by warm water in circulating pipes; for lengths of more than 800 ft., air calorifers are used, about 800 ft. apart, with ventilating pipes specially heated, to cause the air to circulate. These galleries are usually built below the ground level, or if built above the ground, are banked in. In winter, the rivers are frozen solid, except the Selenga, Uda, Chita, Argun, and Onon. Across the River Khilok, at 484 miles from Irkutsk, a dam has been built; the sluices are closed at the commencement of the frost, and thus a large pond of solid ice is formed. The ice is then cut away in blocks, and thawed by steam in a tank. Water is obtained in this way from December to the middle of March.

Manchurian Sections.—After crossing the Manchurian border near Manchuria station, the line runs through a flat steppe district, as far as the Great Khingan Mountains, the slopes of which it takes near Myandukhe at 1,133¼ miles from Irkutsk (189 miles from Manchuria). The slopes of the Great Khingan are richly wooded. The summit of the pass is penetrated by a tunnel at 1,176 miles, at a height of 3,355 ft. above sea-level. The tunnel, which is 3,383⅓ yds. long, is cut through sandstone, not bricked, and is very wet. There is an old deviation, used when the tunnel was being built, 12½ miles long, with steep gradients. After Barim, at 1,229 miles from Irkutsk, the steppe country begins again and is more fertile than the country to the west of the mountains. From Tsitsihar to Kharbin, the line passes over a plateau without trees and almost without inhabitants. After Kharbin, which is in a marshy district, the line goes through mountainous country all the way to Vladivostok. There are three tunnels about 1,761 miles from Irkutsk. At Pogranichnaya is the terminus of the Chinese Eastern Railway. Between this station and Grodekovo on the Ussuri Railway there are six tunnels. Before reaching Vladivostok there is a deep cutting. Between the stations of Kiparisov, 36 miles from Vladivostok, and Nadezhdinskaya, 28 miles, the permanent way is to be changed and a tunnel 2,450 ft. cut. From Nikolsk-Ussuriski to Vladivostok the line is double. The grades on the Chinese Eastern Railway are in the hill sections 1 in 57 on straights and 1 in 66 on curves. The curves are 1,400 ft., with a few of 840. On the plain, the maximum grade is 1 in 125, compensated to 1 in 165 on curves 2,100 ft. By the treaty of Portsmouth (1905) this line may not be used for military purposes. In 1939 China has the right of buying it.

It was reported in 1901 that Russians had begun the construction of a secret standard gauge railway from the Chinese Eastern Railway, about 70 miles west of Khailar, southward through Manchuria. The railway was said to have been built for about 300 miles by 1902. This statement was proved to be untrue. The report possibly originated in a line to some quarry laid during the construction of the Chinese Eastern Railway.

The Amur Railway

The Amur Railway consists of a portion of the old Karimskaya–Stryetensk line, as far as Kuenga, to which point it follows, first the Ingoda valley, and then the Shilka, which is formed by the confluence of the rivers Ingoda and Onon, near the station of Onon. The Shilka, like the Ingoda, is winding, with steep banks. The line follows its left bank along a ledge hewn out of the rock. At Kuenga the line branches north up the right bank of the River Kuenga (which it crosses near Ukurei), and then north-east. It goes through a hilly and wooded country, little peopled, more or less parallel to the Shilka valley at distances from it varying from 16 to 80 miles as the crow flies and out of artillery range from the Chinese side of the Amur. There is a bad road along the north bank of the Shilka, and a road is being made along the railway. The embankments and bridge supports are built for two tracks although the present line is single. It is said to be capable of taking nine trains a day each way with ease, to have few tunnels, and that normally the maximum grade provided for is 1 in 100 with minimum curves of 1,050 ft. On the spurs of the Great Khingan and Lagar-Aul the grade is 1 in 71 with curves of 820 ft. The head-quarters of the Amur Railway are at Aleksyeevsk on the River Zeya. All the artificial work in connexion with the line is to be of a permanent nature; the weight of the rails is 22 lb. per running foot. The permanent way is badly laid, and floods have done the line great damage, so that by the middle of 1917 it was being used comparatively little.

The Amur Railway is a reversion to the original course projected for the railway to the Pacific, and it has been necessary to return to the adoption of this course now that the Manchurian Railway is no longer under Russian control since the war with Japan. The construction of the first section (Kuenga to Uryum, 122 miles) was begun in 1908, and continued through the next two years, traffic being partially started in the winter of 1910. In 1912 two passenger trains ran in this section every 24 hours. The next section from Uryum to Kerak (398 miles) was commenced in 1910, the third section from Kerak to the River Diya near the Bureya (435 miles) in the spring of 1911, the last section to Khabarovsk (302 miles) in the spring of 1912. The central portion was completed in 1914, and the whole, including the great bridge over the Amur, the longest in Russia, in 1916. The work was to be done exclusively by white labour with no help from the yellow races. The workmen were to be mainly reservists who it was hoped would, on the completion of the line, form military colonies in the Amur and Primorsk territories. The difficulties in construction were greater than in any portion of the Siberian Railway. The line ran through an almost uninhabited district, to which roads had to be made, and the difficulties of water supply were enormous, because the rivers froze in winter, and such supply as there was could only be procured by excavating to sources below the river-bed which did not freeze. The marshiness of the ground in the central section added to the difficulty. Many parts of the route were only to be traversed on horseback or on foot, and during the thaw and in the rainy season from June to August were entirely inaccessible. The country opened up by this railway is being rapidly populated: big villages are growing up where the line crosses the Zeya and the Bureya, the village of Surazhevka having been made already into the town of Aleksyeevsk. The western part of the line abounds in gold and there are coal deposits on the lower reaches of the Bureya. It is ascertained that 300,000 colonists can be suitably settled in the district east of the Zeya; twenty industrial settlements are growing up along the line, and a successful future for the timber industry is anticipated.

The Ussuri Railway

At Khabarovsk the Amur Railway joins the Ussuri Railway. The line follows the right bank of the Ussuri River, crossing frequent tributaries and high watersheds formed by the outspurs of the Sikhota Alin Range. Beyond Kruglikov the country becomes flat and marshy again, and then the line passes through virgin forest of gigantic cedar and larch, intertwined with wild vines and creepers. At Gedik the highest point on the line is reached, 445 ft. above sea-level. To Muravievo-Amurskaya the line continues to follow the right bank of the Ussuri, of which the left bank belongs to Manchuria. The country is hilly and wooded. The line is from 2 to 26½ miles from the river. After crossing the Ussuri, the line enters the swampy Prikhankoisk basin along Lake Khanka, after which it crosses the watershed between the Lefu and Suifun Rivers. At Nikolsk-Ussuriski the line joins the main railroad to Vladivostok. The double track between these two places is now completed. The grades are as follows: for the first 228 miles from Khabarovsk (i. e. to 2 miles before Ussuri), 1 in 100 in curves up to 3,500 ft., and 1 in 125 in curves of 2,100 ft.; from there to Nikolsk-Ussuriski the grade is 1 in 125 in 2,100 ft. curves.

The Altai Railway

The railway from Novo-Nikolaevsk to Barnaul and Semipalatinsk was opened in November 1915. It opens up an important mining district, and the advisability of constructing such a line was demonstrated by the fact that from its opening the number of passengers was much larger than the builders anticipated. The line could not carry all the passengers, and many were compelled to wait several days. Daily trains with cars of 3 classes were running regularly, the cars being new and comfortable. When the line was first used the station-buildings were unfinished, and a dining-room only was opened in Barnaul. A branch line runs from Altaiskaya to Biisk.

Other Siberian Railways

Three other important branches from the main line of the Siberian Railway have recently been opened.

From Tatarskaya, 105 miles east of Omsk, a line goes south-east into the Kulundinsk steppes to Slavgorod. It was built by private enterprise. The whole region it traverses was practically uninhabited in 1907 but by 1912 was being rapidly settled by emigrants. It is proposed to extend this line to Semipalatinsk, and to link Slavgorod to Pavlodar on the River Irtish, and to Barnaul on the River Ob and the Altai Railway.

From Yurga, 385 miles west of Krasnoyarsk, there is a short line, via Kolchugino, built by private enterprise to tap the coal-fields lying north of Kuznetsk, which town is the terminus of the railway.

The Minusinsk line leaves the Siberian Railway at Achinsk and extends through hilly country to Minusinsk on the River Yenisei. It passes through a fertile region with considerable coal and iron resources.

Three short railways built for mining purposes are not connected with the main Siberian Railway. One is between the coal mines of Ekibas-tusc and the River Irtish. It is 70 miles long and is of standard gauge. A second is from the zinc mines of Riderski to Ust-Kamenogorsk. It is 70 miles long and has a 3 ft. gauge. The last runs from Bodaibo to the Vitim, a distance of 15 miles. It is narrow gauge.

The Murman Railway

The total cost of construction is estimated at £2,200,000. From Zvanka, 76 miles from Petrograd, a line has been constructed by the Olonets Railway for 176½ miles to Petrozavodsk. This has been built with French capital but taken over by the Russian Government. From Petrozavodsk to Murmansk the line is divided into 3 sections: (1) from Petrozavodsk to Soroka where the line reaches the White Sea, opened in 1915, (2) from Soroka to Kandalaksha, opened in 1917, (3) from Kandalaksha to Murmansk, opened in 1916. The last section was constructed by British engineers. Traffic had started on the third section before the middle section was completed, but, while it was being made, traffic was taken from Kandalaksha to Kem or Soroka by sea. The whole railway is of Russian standard gauge and is the property of the Russian Government. The line is single and there are passing sidings every 10–12 miles. The line will afford opportunities for exploiting timber and wood-pulp. At present its chief importance is strategic.

Between Petrozavodsk and Kem the line traverses wild and almost uninhabited country, full of forests and lakes. From Kem to Kandalaksha it passes through a succession of small lakes, swamps, and virgin forests. The third section goes across the Kola Peninsula, which is uninhabited but for the coastal strip. This part of the line, however, was of comparatively easy construction, and the material for it was shipped direct to Kola.

The section between Murmansk and Kem becomes almost unworkable in summer owing to the swampy nature of the country. No great speed can be obtained. The rest of the line is firmer.

Probably six trains a day could pass each way in case of necessity. There are two patterns of trucks, the Russian and the American. Large numbers of the latter are being used. The closed trucks have the following dimensions :

	Russian.	*American.*
Length	21 ft.	13 ft.
Breadth	9 ft.	8 ft. 9 in.
Height	7 ft. 3 in.	7 ft. 9 in.
Carrying capacity	20 tons	39 tons

Arkhangel–Vologda Railway

This line was narrow gauge at the outbreak of the war. To cope with the increased traffic via Arkhangel, a third rail for standard-gauge rolling-stock was laid from Arkhangel to Nyandoma (214 miles). Thence to Vologda (180 miles) a standard-gauge line was laid beside the narrow-gauge line. Later the narrow-gauge line between Nyandoma and Vologda was converted to broad gauge, thus giving a double track on this part of the railway. It was then proposed to double the line from Arkhangel to Nyandoma.

In 1916 the maximum number of trucks that could be dealt with in a day from Arkhangel was 475.

New and Projected Railways

The most important schemes of railway-construction are :

(1) The Southern Siberian Railway, which is under construction, will link up the Altai Railway from its present terminus at Semipalatinsk to Orsk, passing by Akmolinsk and Atbasar. From Orsk it is to be prolonged to unite with the Tashkent line at Martuk Station. It is anticipated this line will greatly stimulate emigration from the less productive provinces of European Russia into lands where there are millions of suitable acres waiting for development, but at present too far from any railway for colonization to be possible.

(2) The railway already constructed from Poletaevo to Troitsk will be continued to Fedorova and Kustanai in the northern part of the Turgai Province.

(3) A line from Petropavlovsk to Kokchetav in the north of the Kirghiz steppe has been mooted. It will be about 100 miles long and will serve the country between the Ishim and the Irtish.

(4) A line was opened on January 11, 1917, from Yekaterinburg by way of Irbit and Turinsk to Saitkovo on the Tavda; it is proposed to continue it to Tobolsk.

(5) A very important line has been projected from Soroka on the Pomorski coast to the River Ob, a distance of about 1,000 miles. It was originally intended to have the terminus at Arkhangel, but this idea seems to have been given up in favour of a terminus at Soroka, where a harbour for vessels of larger draught can be constructed. Such a railway is of immense importance in order to bring about the commercial success of the Murman Railway. The route presents no great difficulties of construction, and would not entail the climatic difficulties that attended the construction of the Murman Railway. In 1916 both routes (to Arkhangel and to Soroka) were being surveyed. The line, as originally proposed, was to run from Arkhangel to Pinega and then through the Ukhta district, and over the Pechora River at Troitsko-Pechorskoe to a landing-stage on the River Ob near Chemashevskoe (63° N. lat.). There is to be a branch on the east side of the Urals to the Nadezhdinski works on the Bogoslovski Railway linking up with Yekaterinburg. It is anticipated that the cost will be £10,032,000; it is estimated that the annual goods traffic over the line will be from 30 to 45 million cwt. The objects of the line are to develop the forest regions of northern Russia, to stimulate the mining of the northern Urals, and to provide Siberia with an important outlet for its trade. The length of the line will be about 1,000 miles. Another proposal, which has more to recommend it, is to extend the Vyatka-Kotlas line north-west to meet the Murman Railway at Soroka on the White Sea.

(6) Closely related to these schemes is the project for the so-called Arctic Railway from Obdorsk, or some place on the River Ob near by, across the Ural Mountains to a port in the

eastern Barents Sea. This would obviate the difficult passage of the Kara Sea and make a good outlet for the trade of Siberia. The Urals afford fairly easy passes for such a railway. The former proposal for a port on the Gulf of Baidaratskaya has been abandoned in favour of a port near the mouth of the River Pechora, possibly near the entrance to Khaipudirskaya Bay. The length of the line will be about 300 miles. Narrow gauge is proposed. The cost, exclusive of harbour works, is estimated at about £2,250,000.

(7) Other proposals are a line from Achinsk to Yeniseisk, a trunk line from Tyumen to Tomsk, crossing the River Irtish at Tara, and a line north to Tara from the existing railway lines. Early in 1917 it was decided to accelerate the construction of a line from Yaroslav in European Russia by way of Kostroma, Krasnoufimsk, Ufalei, and Ishim to Tomsk.

(8) It is proposed to connect the gold-mining district of the Lena with the Siberian Railway by a line from Tulun which will reach the Lena valley at Ust-Kutskoe (94 miles). The inhabitants of Irkutsk are anxious that the line should go from Irkutsk via Zhigalovskaya and for this reason are agitating to improve the river-bed between Ust-Kutskoe and Zhigalovskaya. It is further proposed to carry the line on to Bodaibo on the Vitim via Kunerma, the distance being 1,000 miles from Tulun.

(9) There is talk of a railway either from Misovaya or from Verkhne-Udinsk to Kyakhta in order to bring the Mongolian markets into connexion with the main Siberian line. This railway is to be constructed at government expense, and had been approved by the Council of Ministers in 1913. It is understood that it was arranged in 1915 to extend this line to Urga, because the fall of Tsingtau and the expulsion of German trade from nearly all the far east has given Russia great opportunities of acquiring fresh markets in China.

(10) An agreement was signed on March 27, 1916 for a new railway to be constructed connecting the Chinese Eastern Railway at Tsitsihar with Aigun on the Amur, and continuing a few miles along that river so as to be opposite Blagovyesh-

chensk. A branch line will run from Mergen about halfway along this route to Kharbin, which will increase the agricultural importance of this district; farther north timber and mining are more important industries. The length of these lines will be about 650 miles in all.

(11) A short line is projected from de Castries Bay to Sofiisk on the lower Amur.

(12) A line was strongly advocated by the late Governor-General of the Priamur to link the Ussuri Railway with Olgi Bay and Imperatorskaya Harbour, which would be of great service for the export of timber.

(14) According to the Russian press the committee under the chairmanship of the Assistant Minister of Ways and Communications included in the estimates of 1917 the execution of an economical and technical survey of the railroad from Aleksyeevsk to Nikolaevsk on the Amur.

CHAPTER XVIII

HISTORICAL NOTES[1]

Prehistoric Races—Early Relations with Russia—The Early Conquerors—Attempts to conquer the Amur Region—The Treaty of Aigun—The Peking Convention—Russia and Japan—Russian Advance in Mongolia.

In prehistoric times Siberia, especially in the south-west, was far more densely populated than it is at present. Neolithic remains are numerous. Many peoples have doubtless been driven into the inhospitable north before successive waves of emigration.

Prehistoric Races

In the Yenisei valley and throughout the south-west of Siberia there are found a number of tumuli containing the remains of a highly-developed bronze civilization, when gold and silver were also largely worked. These are generally thought to belong to a primitive Yeniseian race. In the third century the Uigurs, a Turkic stock, are said to have overrun the country during the time of the wanderings of the Huns, and the resulting people are generally known as Ugro-Samoyedes. They built tumuli over their dead, which they adorned with monoliths. These were sometimes simple pillars, but were more often carved into human likenesses, many of them with astonishingly realistic features. These monoliths occur throughout south-western Siberia, but are most common in the Minusinsk district, where they invariably face north and south. Iron and bronze implements are found in large numbers in the tombs, but gold and silver are rare. The Ugro-Samoyedes were great agriculturists, irrigating

[1] These notes are not carried beyond 1916 and so do not include the Russian Revolution of 1917.

wide tracts of land. Modern settlers not infrequently open up and use their canals to-day. Eight centuries later these Ugro-Samoyedes were subdued by another Turkic stock, also highly civilized, which maintained its power till the Mongols under Jenghiz Khan swept over the land and utterly destroyed its civilization.

Early Relations with Russia

By the end of the eleventh century the energetic merchants of Novgorod had penetrated into Siberia, or Ugra, as it was then called, as far as the present government of Tobolsk. In the fourteenth century they even established settlements on the Taz. These flourished until the Russian Government closed the Kara Sea. Not only Russian, but Dutch and English merchants, and even Russian emigrants, used the Kara Sea route at this time until embargoes were placed upon it. But in 1662 the Ostyaks destroyed the rapidly-decaying settlements on the Taz.

These men of Novgorod were mere fur-traders, with no idea of conquest. But meanwhile the people of Moscow were systematically advancing towards the Urals, and during the sixteenth century began to enter into close relations with the tribes on the other side. Tartar hordes had recently brought them into some kind of subjection. In 1555 Ediger Khan, who had united the small Tartar principalities into a kingdom, consented to pay a tribute of 1,000 sables to Moscow in return for a protection which the Tsar was incapable of affording, though he welcomed the tribute.

The Cossacks

The conquest of Siberia was the work of the Cossacks. These are not, as is generally supposed, a body of irregular horse, but a section of the Russian people with special duties and special privileges. Thus in 1851 Count Muraviev converted the Nerchinsk peasants into Cossacks. The Cossack unit is the *stanitsa* or village. They hold their land in common and have the right to let it. They are liable to military service between the ages of 18 and 48. They receive a money-allowance from the Government, which also supplies them with arms, but they must provide their own equipment and their own horses, if mounted. Large stretches of land are reserved for them, usually on the frontiers. They are divided into *sotnyas*, each of which mannages its own affairs.

The Early Conquerors

Yermak, the first conqueror of Siberia, was a tracker on the Volga, then a pirate on that river among the Don Cossacks, till his success attracted the dangerous attention of the authorities. He fled to the Stroganovs, the great merchant family of Perm. They had long coveted the rich furs of Ugra and were only too glad to make use of Yermak to satisfy their ambitions. They provided him and his 800 men with everything they required for their expedition, including three priests and a runaway monk. In 1580 Yermak reached the Tura, wintering where Tyumen now stands, and in the following year he took the famous fort of Isker, or Sibir, near Tobolsk, from which the name of Siberia is sometimes derived. Ivan the Terrible rewarded him with a free pardon for his early misdeeds. By 1584, when he was drowned in the Irtish, Yermak's conquests extended from the confluence of the Ob and the Irtish to the Tagil and the Tura, and he had secured them by forts at Tyumen and Tobolsk.

The romantic story of the conquest of Siberia recalls that of Mexico or Peru. It was made possible only by the helplessness of the natives in the face of firearms. Yermak's success shook the cohesion of the Tartar power, and his conquests therefore survived his death. A stockaded *ostrog*, garrisoned by a few Cossacks with a gun or two, could hold down an enormous expanse of country in the wild northern regions. But the Cossacks were as yet unable to overthrow the stronger barbarian organizations of the south. Hence they followed the line of least resistance towards the east and the north. Thus Berezov was founded in 1593, 11 years before Tomsk. Furs were to the Cossacks what gold was to the Spaniards, and these were still to be found in plenty in the sparsely inhabited districts into which they penetrated.

Moreover, the Cossacks were born sailors. We find Yermak for instance, damming a stream with sails in order to secure sufficient water for his boats. Their natural method of progress was to sail down one river, haul their boats over the portage, and then sail down another. And none of the principal rivers of Siberia runs south. Yeniseisk was founded in 1618, and 12 years later the Lena was reached. The Yakutsk *ostrog* dates from 1637. *Ostrogs* were also established at Tomsk, Turukhansk, Olekminsk, Irkutsk, and other places. The conquerors met with little effective resistance from the scanty population, though the Tungus were only subdued, after a desperate struggle about 1623.

Access to the Pacific Ocean

As there are no tributaries connecting the Lena with other great eastern rivers by easy portages, the Cossack Buse, who was sent to collect tribute from the northern tribes, sailed down the Lena in 1638 and out at its western arm to the Olenek and the Yana. In 1639 he discovered the Indigirka. The hardships endured by the Cossacks on these expeditions in the unknown Arctic Ocean were often terrible, and, unlike Yermak, they soon began to treat the natives with the utmost cruelty. In 1644 Nizhne-Kolimsk was founded on the Kolima. Four years later the Cossack Dezhnev, after whom the Russians have justly re-named East Cape Cape Dezhneva, made his wonderful voyage round the coast to the Anadir, sailing through the Bering Straits 80 years before Bering re-discovered them. On the Anadir he was joined by other Cossacks, who had come from the Kolima by land along the Anyui and over the watershed. The Okhotsk *ostrog* was built in 1647, in spite of the fierce resistance of the Buryats. By 1697 Kamchatka was explored to Cape Lopatka and an *ostrog* built at Verkhne-Kamchatsk. But there were serious mutinies and quarrels among the

Cossacks and their leaders in this remote district, which was highly prized on account of its wealth in furs.

Attempts to Conquer the Amur Region

In 1643 Poyarkov started up the Aldan and then sailed down the Zeya and the Amur to its mouth, returning by the Sea of Okhotsk to the mouth of the Ulya, then back to Yakutsk by the Maya. His name is commemorated in Poyarkova on the Amur. This remarkable voyage lasted three years.

We now meet with the second great name among the Siberian conquerors, Khabarov, the merchant of Olekminsk, who occupied a position on the Lena not unlike that of the Stroganovs at Perm. In 1649–50 he fitted out an expedition to the Amur, taking the much shorter route along the Olekma. He seized Albazin, then sailed down the Amur and wintered at the place which now bears his name, Khabarovsk, where he successfully resisted the attacks of large Chinese forces. Compelled to retreat, he established a post at the mouth of the Zeya. The fact that he instinctively selected Albazin, Blagovyeshchensk, or Ust-Zeya as it was then called, and Khabarovsk for his posts—all places which have since proved to be of primary importance in the history of the river—is a striking proof of his foresight. He also insisted from the first that at least 6,000 men were necessary for the conquest of the Amur, and he would probably have succeeded in the enterprise, had they been forthcoming. But such an army could not have been raised or even supported in Siberia at this time.

Meanwhile the Russians were advancing in the Transbaikal, the natural base for an attack on the Amur. In 1649 Verkhne-Udinsk was founded, and by 1654 Nerchinsk, on the other side of the watershed. Then Beketov, after whom Beketova on the Amur is named, pushed along the Ingoda and the Shilka till he reached the Amur proper, thus discovering the quickest way to the river. The Cossacks still maintained their hold, but the Chinese, exasperated by their raids along the Ussuri into the heart of Manchuria, were determined to drive them out, at all costs.

In 1655 they failed to take Kumarskaya with an enormous army, and for a time Russian prospects looked brighter. But even Tolbuzin's heroic defence of Albazin in 1684 and 1685 was unavailing. His name is preserved in Tolbuzina, close to Albazin, on the Amur. The numbers of the Cossacks were inadequate for their task, and Russia, weakened by internal troubles, ended by ceding the Amur to China by the Treaty of Nerchinsk (1689).

The eighteenth century was marked by the beginning of the scientific exploration and organized settlement of Siberia. Numerous expeditions were sent to subdue the tribes of the north-east, notably the Chukchee, but they were never more than partially successful.

Muraviev and the Amur

As time went on Russia felt more and more the disadvantage under which she laboured in not having a suitable outlet on the Pacific. Her occupation of the Amur was due almost entirely to the energy of Count Muraviev-Amurski, the third great name in the history of Siberia. In 1849 he sent Nevelski, who has given an alternative name to the Gulf of Tartary, to explore the mouth of the river. He thus learnt that Sakhalin was an island and that the river-mouth was accessible to sea-going ships. In 1850 Nevelski established a port at Nikolaevsk in defiance of all rights. Two years later de Castries Bay and Mariinsk were occupied, while posts were established on Sakhalin. Buse and Korsakov, after whom settlements on the Amur have been named, also played an active part in these proceedings.

But Russia owes her hold on the Amur to the Crimean War, in the course of which a strong force of French and British seamen was defeated by a handful of Cossacks in an attempt to take Petropavlovsk in Kamchatka. Not till 1854 did the Tsar consent to an expedition down the Amur under Muraviev. In the following year a large flotilla was sent, bringing much-needed assistance to the Russian Pacific squadron. The settlement of peasants along the left bank was begun in 1856 with colonies at the mouths of the Kumara, the Zeya and the Sungari and at the entrance to the Little Khingan gorge ; and the process was steadily continued.

The Treaty of Aigun

China, which had never attempted to occupy the left bank of the Amur, accepted the inevitable, and by the Treaty of Aigun, in 1858, ceded to Russia all territory on the left bank of the Amur from the Argun to the sea. The territory on the right bank of the Amur from the Argun to the Ussuri went to China. The territory between the Ussuri and the sea was to remain neutral ground between the two empires, pending a delimitation of the frontiers. The Amur, Sungari and Ussuri were to be open to the navigation of Russian and Chinese vessels, but closed to the vessels of other countries.

The Peking Convention

The Peking Convention of 1860 defined the boundary between Russia and Manchuria as following the Argun, Amur, and Ussuri, and Sungacha, to Lake Khanka. From the source of the Sungacha the boundary crosses Lake Khanka to the mouth of the Pai-ling and thence follows a line described in the Convention (Chinese text) as follows : 'From the mouth of the Pai-ling River along a mountain range to the mouth of the Hu-pu-tu River, and from the mouth of the Hu-pu-tu River down the Hunchun River and along the range of mountains between that river and the sea to the mouth of the Tumen River.' 'The frontier meets the Tumen River at about 20 li (6 miles) from its mouth.'

(On all Russian maps the frontier follows the Hu-pu-tu River (Khubtu) throughout its course, and goes along the mountain range between the Hun-chun River and the sea until it meets the River Tumen.)

Russian Ports on the Pacific

Before the days of Muraviev Okhotsk had been the Russian military and naval port on the Pacific, but in 1850, deceived by the beauties of Avacha Bay in Kamchatka in summer, the Governor General had made it the head-quarters of the Pacific squadron. Ten years later the squadron was transferred to Nikolaevsk. Russia was now in possession of Vladivostok, but she did not make it her naval base till 1872, since when its growth has been very rapid.

The Fifty-Verst Zone

The Convention of Peking stipulated that there should be free trade between the two empires along the new frontiers. An agreement extending over a period of 30 years, which was confirmed in Petrograd in 1881, established a 50-verst zone along the entire frontier within which no customs dues were to be collected. In practice this meant that Chinese goods were admitted duty-free into Siberia along the Amur, though Russian merchants also did a considerable trade with China further west. Russia denounced the agreement on its expiration on January 1, 1912.

Russia and Japan

Manchuria and the Russo-Japanese War

Russia began her Manchurian adventure with the agreement between the Chinese Government and the Russo-Chinese Bank in 1896. This brought into being the Chinese Eastern Railway Company, which was to link the Transbaikal Railway with the Ussuri Railway at Vladivostok. Shares could only be held by Russians and Chinese. Russia thus definitely abandoned the longer and more difficult route along the Amur through her own territory in favour of the shorter route through Manchuria. The railway company had the right to a strip of territory on each side of the line within which it exercised absolute control, while Kharbin in the Kirin Province was established as the head-quarters of the line. It has now become one of the most important Russian towns in the Far East.

Two years later, in 1898, Russia obtained the lease of Port Arthur and Talienwan (Dairen or Dalny) on the Liaotung Peninsula. After the Boxer rising in 1900, when the Chinese were driven into the Amur at Blagovyeshchensk and Aigun was destroyed, the Russians occupied large areas in Manchuria which they declined to surrender. They rapidly extended their influence both in Manchuria and Korea.

Then came the Russo-Japanese War, followed by the Treaty of Portsmouth of 1905. Russia recognized Japan's paramount influence in Korea, ceded to her all her rights in Port Arthur and Talienwan (Dairen) and the railway between Port Arthur and Changchun and handed over to her the occupied districts in Manchuria, which were restored to China. Japan was also entitled to fishing-rights in the Japan and Bering Seas and the Sea of Okhotsk. Japan had acquired the Kuril Islands in exchange for the southern portion of Sakhalin by a treaty with Russia in 1875. But the Treaty of Portsmouth again divided the island between the two powers at the line of 50° N. lat. Russia thus abandoned her attempts to become the dominant power in Manchuria, where her authority has rapidly declined, and the Chinese themselves are now steadily settling the province. The construction of the Amur Railway thus became a strategic necessity for Russia. The building of a line from Aigun southward to Tsitsihar, linking the Chinese Eastern and Amur Railways, could only be a question of time, and in 1916 an agreement was concluded between Russia and China by which Russia is to raise a 5 per cent. loan of £5,000,000 after the war for the building of this line. It is to be controlled by Russians nominated by the Russo-Chinese Bank.

Russo-Japanese Agreement of 1916

The growing cordiality of the relations between Russia and Japan, which resulted in the alliance of 1916, is further exemplified by Russia's agreement to sell Japan the southern half of the Kharbin-Changchun Railway, comprising a section of about 60 miles and including the line from Changchun to the left bank of the Sungari.

Japan has never accepted Russia's claim to exclude all

other foreigners from the navigation of the Sungari in accordance with the Treaty of Aigun, and Russia now recognizes the right of Japanese shipping to navigate the Sungari between Kirin and Petuna.

Russian Advance in Mongolia

The establishment of the 50-verst zone in 1881 gave a great stimulus to Russian peaceful penetration into Chinese territory south of the Yeniseisk Government. Several Russian towns, of which Turanski is the chief, have sprung up over the border and Russian settlers are increasing. The natives prefer to resort to the Russian schools and courts, where they are better treated. Similarly the Russian merchants are said not to exploit them to nearly the same extent as the Chinese, who profit by their vices. But the success of the process depends on the tact of the officials. Since the war they are said to have become more autocratic. Hence the natives are growing discontented and retiring further south.

Meanwhile the Chinese are steadily colonizing Mongolia. This movement, like their immigration into Siberia, is probably economic in origin, whatever results it may ultimately bring about. It has, however, caused great dissatisfaction among the nomad Mongols who have also suffered from the Chinese methods of trading. On the overthrow of the Manchu dynasty in 1912 the Mongol chiefs claimed their independence, holding that they owed no allegiance to its successors.

A Russo-Mongolian agreement was signed on Oct. 21, 1912, recognizing the autonomy of Mongolia, but it was challenged by China, who insisted on her sovereign rights over Mongolia. In 1913 Russia admitted China's claim to suzerainty in Mongolia on condition that Mongolian autonomy was recognized. By this agreement Russia secured for herself substantial privileges in that country. The agreement does not appear to have been received with much favour by the Mongols.

APPENDIX

Weights and Measures

I

Measures of weight

Russian.	*English.*	*Metric.*
1 dolya	0·68 grains	4·44 centigrams
1 zolotnik (96 dol.)	0·15 oz. av.	4·26 grams
1 lot (3 zol.)	0·45 oz. av.	12·79 grams
1 funt (96 zol.)	0·902 lb.	0·409 kilograms
1 pud (40 funts)	0·32 cwt. (36·11 lb. av.)	16·38 kilograms
1 berkovets (10 puds)	3·22 cwt.	163·804 kilograms

Measures of length

1 dyuim	1 in.	25·4 millimetres
1 fut	1 foot	304·8 millimetres
1 vershok	1·75 in.	4·44 centimetres
1 arshin (16 v.)	2 ft. 4 in.	0·71 metres
1 sazhen (3 ar.)	7 ft. (in liquid depth 1 fathom)	2·13 metres
1 verst (500 sazh.)	3,500 ft., or 0·6628 mile	1·06 kilometres

Measures of area

1 sq. arshin	5·44 sq. ft.	
1 sq. sazhen	5·44 sq. yd. (49 sq. ft.)	4·54 sq. metres
1 desyatin (2,400 sq. sazhens)	2·7 acres	1·09 hectares
1 sq. verst (250,000 sq. sazh.)	281·221 acres	113·80 hectares

Measures of volume

1 cubic vershok	5·35 c. in.	87·81 c.c.
1 cubic arshin	0·47 c. yd.	0·35 c. metre
1 cubic sazhen	12·70 c. yd.	9·71 c. metres

Liquid measures

1 charka	0·21 pints	0·12 litres
1 bottle (5 ch.)	1·08 pints	0·61 litres
1 shtof (10 ch.)	1·08 quarts	1·22 litres
1 vedro (100 ch.)	2·706 gallons	12·29 litres
1 bochka (40 ved.)	108·27 gallons	4·91 hectolitres

Dry measures

1 garnets	2·88 gallons	3·27 litres
1 chetverik (8 g.)	0·72 bush.	26·23 litres
1 chetvert (8 chk.)	5·77 bush.	2·09 hectolitres

II

Measures of weight.

English.	*Russian.*
1 oz.	6·64 zolotniks
1 lb.	1·107 funt
1 cwt.	3·104 puds
1 ton	62·02 puds

Measures of length

1 in.	0·57 vershok
1 yd.	0·42 sazh. or 1·28 arsh.
1 mile	1·508 versts or 754·28 sazh.

Measures of area

1 sq. yd.	0·18 sq. sazh.
1 acre	888·97 sq. sazh. or 0·37 desyatin
1 sq. mile	2·27 sq. versts or 237·06 desyatins

Measures of volume

1 cubic yd.	0·078 cubic sazh.

Liquid measures

1 pint	4·61 charka
1 quart	9·23 charka
1 gallon	0·36 vedro

Dry measures

1 gallon	0·34 garnets
1 bushel	1·38 chetverik
1 quarter	11·08 chetverik
1 chaldron	6·23 chetvert

It is reported that the Russian Government has decided to introduce the metric system in August 1921 and to prohibit the use of the old system of weights and measures from January 1, 1925.

MONEY

The legal unit is the silver rouble of 100 kopeks. It is treated usually as the equivalent of 2*s.* 1*d.* in our currency, but the exchange is very variable. In this book for all large sums and round numbers 2*s.* has been taken as the equivalent; in official calculations 9·46 roubles are taken as equal to the pound sterling. Gold coins are the imperial and half imperial of 15 and 7½ roubles. New gold coins are issued bearing the inscription of 10 roubles and 5 roubles. Besides the silver rouble, credit notes (500, 100, 50, 25, 10, 5, 3, and 1 rouble) are legal tender. Paper money of all denominations down to one kopek has been in use since the outbreak of the war.

TIME

Local mean times are used throughout Siberia. Irkutsk time is 6 hrs. 57 min. 15 sec. fast, and Vladivostok time 8 hrs. 47 min. 34·5 sec. fast on Greenwich mean time. Russia in Europe uses Petrograd time, which is that of Pulkova Observatory, 2 hrs. 1 min. 18·7 sec. fast on Greenwich mean time.

Calendar

The Julian calendar is still in use in Russian lands. It is 13 days behind the Gregorian calendar used in other countries, Throughout this book, unless old style (O.S.) has been stated, all dates are given in the Gregorian calendar.

It was reported in 1918 that the Gregorian calendar was to be adopted throughout Russian lands.

GLOSSARY

This glossary contains Russian and other words frequently used in the text, including words, with their usual abbreviations, which occur on Russian maps. In the case of adjectives the masculine termination (*-i*) is given. The neuter termination is generally *-oe*, and the feminine *-aya*.

Aba, tomb of local hero (Kirghiz).
Artel, group of workers ; trade union.
Aul, encampment.
Balagan, hut of wood and thatch.
Balog, family sledge.
Boloto, marsh.
Bolshoi, Bol., great.
Bor, hill, sometimes pine forest.
Brat, brother.
Brodyagi, escaped convicts.
Byeli, white.
Cherni, black.
Chernozem, black earth.
Chud, primitive inhabitants of Siberia.
Chum, Samoyede or Ostyak tent.
Dekabrists, those who took part in the plot of Dec. 14, 1825.
Derevnya, village without church.
Dobrovolni, voluntary followers of exiles.
Doroga, road.
Drozhki, cab.
Dukhobors, sect, 'spiritual fighters.'
Duma, council or council chamber.
Dur, dying of the water on the Ob, &c., see Vol. II. Chap. V.
Fabrika, factory.
Forpost, F., military outpost.
Gora, G., mountain.
Gorbusha, kind of salmon.
Gorod, town.
Guba, bay.

Guberniya, government.
Ispravnik, official in charge of district; police commissioner.
Izba, winter hut.
Kamen, rock; stone, or cliff.
Karaul, Kar., picket-station; guard-house.
Kayak, skin boat for one man.
Kedr, white cedar or cembra pine.
Kereoshka, boat-sledge drawn by reindeer.
Kerka, log-house of Zirians.
Keta, kind of salmon.
Khodok, advance agent of emigrants.
Khrebet, mountain range.
Kolodnik, convict.
Kozha, leather; a Lapp name for seals.
Krasni, red.
Kul, lake.
Kumis, fermented mare's milk.
Kurgan, tumulus; burial mound.
Kuropatka, willow-grouse.
Kvas, intoxicant made from barley.
Lyeto, summer.
Mali, Mal., small.
Maralnik, farm where maral deer are kept.
Mir, village assembly.
Mis, cape; headland.
Mogila, Mog., tomb.
More, sea.
Navaga, kind of cod.
Nizhni, lower.
Nos, headland.
Novi, new.
Nyelma, kind of salmon.
Oblast, territory.
Ostrog, block-house; stockade.
Ostrova, island.
Otrassl, Otr., mining settlement in Urals
Ozero, lake.

Pagost, Lapp settlement.
Pereket, Per., sand-bank or bar.
Pereval, Per., pass.
Perevoz, ferry.
Piket, Pik., P., picket station.
Pless, straight reaches of a winding river.
Pochtovaya kontora, post-office.
Poch. stanitsa, posting station.
Pogost, church without village.
Pogron, government tax.
Polustantsia, intermediate station or stage.
Porog, rapid on a river.
Posad, Pos., suburb.
Povarnya, Povarni, post-station on remote road.
Poviet, storeshed.
Pristan, Pr., landing-place.
Proliv, strait.
Raskolnik, dissenter.
Rasputitsa, season impossible for travel.
Razdvoenie, railway junction or crossing.
Razyezdni puti, railway siding.
Ryeka, river.
Samovolni, voluntary emigrants.
Sast, town dweller of Turkish origin.
Selo, village with church.
Shar, strait.
Sklad, Skl., warehouse or depôt.
Skoptsi, eunuchs, a fanatical sect.
Sori, shallow backwater.
Sredni, middle.
Stanitsa, Stan., station ; Cossack post.
Stari, old.
Starovyeri, old believers.
Svyatoi, holy.
Syeverni, north.
Taiga, coniferous forest.
Tarantass, four-wheeled carriage.

Tarine, freezing to the bottom of certain rivers.
Telega, cart.
Tolsti, thick.
Trakt, great Siberian road.
Treska, cod.
Troika, team of horses.
Tundra, swampy, treeless Arctic plains.
Tupa, winter dwelling of Lapps.
Uba, tomb of local hero (Kirghiz).
Urman, swampy thickets.
Urus, Yukaghir tent.
Uste, Ust-, river mouth.
Uyezd, district.
Verkhni, upper.
Viezha, summer dwelling of Lapps.
Viski, channel between lakes.
Volok, isthmus ; portage.
Volost, canton or cantonal assembly.
Vostok, east.
Yamshchik, posting driver.
Yar, cliff ; bluff.
Yarus, great lines used in fishing.
Yassak, tribute.
Yug, south.
Yurta, Yurt, tent.
Zaimishche, Zaim., low ground between river and hills.
Zaimka, Z., settlement of one or a few houses.
Zaliv, strait.
Zamor, the dying of the waters, see *dur.*
Zapad, west.
Zavod, factory.
Zemlya, land.
Zemstvo, provincial assembly.
Zhelyeznaya doroga, Zhel., railway.
Zhelyezoplavilni zavod, Zhel., iron foundry.
Zherlo, mouth.
Zimia, winter.
Zimove, winter dwelling.

INDEX

Spotted something that isn't quite right?

Let us know and we'll fix it for future reprints. And as a thank you for your diligence, we'll send you a free Sothis book of your choice.

publisher@sothis.press

This page intentionally left blank

Except of course for that line above. And this one.
And whilst we have your attention, perhaps you'd
like to follow us on Instagram @sothis.press. Go on.

Made in United States
Cleveland, OH
05 November 2026

47772343R00227